Unmasking Maskirovka:
Russia's Cyber Influence Operations

Daniel P. Bagge

Defense
Press

Disclaimer

The views expressed in this book are those of the author and do not necessarily reflect the official policy or position of the National Cyber Security Center of the Czech Republic, the National Cyber and Information Security Agency of the Czech Republic, or the Government of the Czech Republic. The author of the publication enjoys full academic freedom, provided he does not disclose classified information, jeopardize operations security, or misrepresent official Czech policy. Such academic freedom empowers him to offer new and sometimes controversial perspectives in the interest of furthering debate on key issues.

First published 2019
by Defense Press
New York, New York

Library of Congress Cataloging-in-Publication Data
A catalog record has been requested for this book.

ISBN: 978-0-578-45142-8

CONTENTS

FOREWORD

The news of foreign manipulation in the United States electoral process highlights an adversary's willingness and capability to attack on a virtual front to achieve geo-political objectives. Revelations of cyber espionage, manipulation, and digital disinformation campaigns underscore the bold new contested spaces in which nations, corporations, and individuals now operate.

Dan Bagge's *Unmasking Maskirovka* is a must read for policy makers and strategists to gain a better understanding of capabilities and limitations in the Information Operations domain. It should be required reading for those working in the information and cyber realms. Dan has put forth an excellent primer in understanding Russia's incorporation of digital capabilities into Information Operations doctrine, the evolution and thought process that feeds it, and how Information Warfare supports Russia's political and military objectives from the strategic to tactical level. With examples stretching from Crimea to the U.S. homeland, Dan's writing is as captivating as it is informative. While most in this domain are simply admiring the problem, Dan offers thoughtful and potentially effective counter-measures.

I first met Dan at the George C. Marshall European Center for Security Studies in Germany, where I was assigned as a Fellow and Adjunct Professor and he was pursuing his post-graduate degree. Dan already displayed a keen interest in cyberspace and information operations and forward thinking in considering the explosion of social media as a new frontier for the extension of adversary policy. He has gone on since to conduct extensive research through subsequent assignments within the security sector, academic programs, and think tanks. As the former Secretary of the Cyber Security Council for the Czech Republic, Dan is one of the foremost advocates shaping not only the defense of his nation in this front, but the dialogue and understanding of the domain across the West.

George Thiebes, COL, USA
Stuttgart, October 1 2018
George Thiebes is a 1990 graduate of the US Military Academy and originally hails from Florida. He currently serves as the J3 at U.S. Special Operations Command Europe. The views presented are those of the author and do not necessarily represent the views of the Department of Defense or its components.

*   *   *

In reaction to Russia´s aggressive behavior since the annexation of Crimea, NATO has reinforced its eastern flank and strengthened its deterrence and defense posture. As the credibility of the article V guarantee is now undisputed, Moscow has accelerated its hybrid activities against NATO, EU, our partners and the West at large. They are staged below the red line level of collective defense. Cyber space is the main battlefield, cyber capabilities are the critical enabler and key force multiplier, while information is the basic weapon.

This book is thus both timely and relevant. The author is a leading Czech cyber defense expert (the team he co-led won the international cyber defense exercise Locked Shields 2017), yet his study focuses primarily on *information warfare*, not cyberwar.

Every day we are exposed to information-based influence operations of many sorts, be it with commercial purposes, such as product and service promotion, or as part of the democratic process and political competition. Digitalized information, the internet, social media and other platforms provide limitless space for information proliferation. In view of this info-overload, the challenge starts with the ability to identify malicious content with hybrid intentions, subsequently to react rapidly with adequate countermeasures, and, in the longer-term perspective, to continue enhancing resilience.

Knowing your opponent is the first step to developing effective countermeasures. The core argument of Bagge´s study emphasizes the fact, that while Russia has conceptualized *cyber power* holistically (embracing not only its information-technological, but also information-psychological aspect), the U.S./Western approach to the problem/solution tends to be more technologically biased and infrastructure centric, not sufficiently integrating those less tangible, i.e. cognitive and perceptual methods of manipulation, be it in the psychological, emotional or linguistic domain.

The author underlines the historical path-dependence of Russian concepts such as active measure, disinformation, deception and denial (so-called *maskirovka*), which are rooted as deep as in the practices of the Tsarist regime´s secret police at the turn of the 19th and 20th centuries. The tradition has been maintained and further sophisticated by the Soviet regime, to be eventually streamlined into a comprehensive concept of *reflexive control*.

The concept, according to Bagge, "is about modelling decision-making systems, understanding them and eventually disrupting them. The main aim is to influence an adversary into making decisions favorable to you as the deceiver." In its essence, it is "a long-term influence campaign, aimed at feeding the adversary with information with an intended impact on the moral values, psychological state or even decision maker´s character." The ultimate aim of the malevolent information operations is to change our threat perception and political behavior, deepen the already existing frictions and fears, spread uncertainties, sow discord in our nations and across our alliances, and finally, undermine our resolve and capacity to act. The creeping character combined with the impact that is graduated over a long term, make information operations especially dangerous as the target often does not fully realize being under attack.

The author offers a meticulous analysis of the Russian model of hybrid warfare in the information space organized around the *reflexive control* concept, including a detailed scrutiny of the relevant Russian strategic and doctrinal documents. Bagge describes a multidimensional architecture of the Russian hybrid model, encompassing a holistic synergy of political, strategic, operational, tactical levels of engagement and a plethora of malign instruments, methods, which are calibrated and used against various targets and audiences. He illustrates its application on several recent case studies.

It is now well established that Russian hybrid approach knows no distinction between peacetime and war. Policy and politics are a continuation of war by other means – as Lenin put it, turning Clausewitz´ famous dictum on its head. It is also important to understand that Russian policy makers and strategists believe in mirror imaging. Thus they are convinced that we (The West) use similar, if not same, concepts and methods of hybrid against them. (The often-quoted article published by the Russian Chief of the General Staff Gerasimov in 2013 is a case in point.) The opposite is true and Bagge is clear: we must refuse the temptation to mimic Russia in order to counter its hybrid campaigns. Accepting Moscow's game would inevitably undermine our societies´ liberal values, ultimately giving victory to Russia.

The study is capped with thirteen concrete recommendations that are related to Russia, which can nevertheless be used for resilience building against hybrid challenges wherever they may come from. Bagge concludes, that "[t]he critical component of any recommendation is, however, the individual. Individuals are the common denominator of the processes and decision-making (...) If the individual is resilient, well then the activities he participates in are resilient, be it analysis, information processing or decision-making."

This book is definitely worth reading. It should reach a wider audience, i.e. politicians and policy makers, experts, journalists, teachers and students, and the interested public. It broadens the perspective of the issue, providing both a thorough analytical framework of the information warfare and a practical manual on how to cope with such threat; it educates and, ultimately, it contributes to the counter-hybrid resilience building.

Ambassador Jiří Šedivý
Brussels, 12 September 2018
Permanent representative of the Czech Republic to NATO, former NATO Assistant Secretary General for Defence Policy and Planning and First Deputy Minister of Defence, Czech MoD. A professor of Security Studies lecturing on defense sector transformation, civil-military relations and national strategy making. As an expert, scholar and opinion maker, he played an important role in the Czech Republic's accession to NATO and EU.

* * *

You are about to read a book that is both up-to-date and unique – a book about Russian information warfare. I got to know Daniel Bagge when he was responsible for formulating and implementing cybersecurity policy in the Czech Republic. Daniel is not just a cybersecurity specialist. He is also one of the true pioneers in this field, intellectually inspiring, a team builder, and a passionate patriot with irrepressible initiative. Therefore, this book is no surprise to me. You could hardly find a better-suited author for writing about this topic.

The book is predominantly covering Russian cyber and information warfare from a wider perspective. This in itself would be sufficiently complicated and topical material. The real value, however, lies in the broad context in which the author analyzes and describes this topic with a true comprehensive approach "through the lens of strategic significance." A reader will learn that the text was really written in accordance with the author's statement, that "considering cybersecurity a technical matter has been a luxury that has led us to the unintended consequences of underestimating its importance for any process or entity." You will find yourself reading about history, motives, perceptions and psyche, where one might expect reading about information technology, networks and malware. And that's exactly what we need!

We entered information age some time ago. As more and more information is being created, transferred, processed and stored in cyber space, this domain is a significant part of our wider information environment. It's time to fully realize our dependency on it, as well as its ever-growing significance for national security. Cyber literacy is no more an issue solely for computer geeks. Understanding cyber warfare in a broader information warfare context is necessary for anyone who wants to contribute to national security. Regardless whether you work in the military, the civilian security establishment, diplomacy, intelligence, or academia, you need to have a basic understanding of context. Cyber and wider information warfare must become a common topic at every level of our military and civilian national security education, training and professional development. It doesn't matter whether you work at tactical, operational, or strategic level; there will always be implications for you.

I truly believe that Daniel Bagge's book is exactly what we needed. It shows us that understanding cyber and information warfare is as well about history, culture, psychology, policy, and strategy as it is about computers, networks, and malware. Daniel guides us through this timely, complex topic and shows us the context and history. Therefore, he is helping us "to contextualize what is happening on a much broader scale." It has a great potential to contribute significantly to our cyber and information warfare awareness, especially (but not only) in relation to the Russian Federation.

It was a privilege and joy to read a manuscript of this book. I thank the author for writing such a book. I believe it will enrich many readers.

Karel Řehka
Gdańsk, 16 September 2018
Brigadier General, Czech Army, Deputy Commander NATO Multinational Division North East, previously commanding Czech special operations forces, author of the book *Information War*

(Informační válka) and contributor to several other books.

* * *

Mr. Bagge has done an excellent job of framing and defining some of the top strategy national issues in cyber and information security. His well-researched historical examples, that include a thoughtful overview and modern interpretation of Soviet "reflexive control," are superb. They serve as a noteworthy reminder for serving national cyber officials on the importance of reflection and continuous study, given the dynamic size, scope and prevalence of cyberspace and its effect on citizens in society. I recommend this book to serving national cyber officials that want to gain a better understanding of the growing cross-functional impacts that they will encounter in the performance of their official duties.

Philip Lark
Washington D.C./Garmisch-Partenkirchen,
19 September 2018
Director, Program on Cyber Security, George C. Marshall European Center for Security Studies

FIGURES

PREFACE

It is as true for cyberspace today as it was correct on the terrestrial battlefield two and a half millennia ago for the Chinese military strategist Sun Tzu: Know your enemy and know yourself and you need not fear one hundred battles. This study synthesizes pre-existing knowledge and concepts and their correlation to the topic of cyber warfare. The aim is to provide readers with the contextualization of the strategic significance of *deception*, cyberspace as an enabler of conducting malicious activities, and how to *amplify and accelerate* previously analog information operations. The book will not be an in-depth technical case study of attacks and instead serves as background reading for policy and decision makers with limited time to conduct their research. It is also a primer for cybersecurity and defense specialists to introduce the big picture outside the technical world.

In the sphere of cyberspace warfare today, two worlds are colliding – the world of policy, decision-making, governance and the national security and the technical world of CERTs and CSIRTs. Both worlds struggle to appreciate one another's importance, and to connect seemingly unrelated incidents with operational art and strategic decision-making.

The aim here is to examine perceptions of Russian strategic and military leadership and Russian military thought processes for employing cyber warfare capabilities. By contrasting this with the U.S./European approach, one can understand the differences, and thereby enable a better understanding of the strategic goals behind Russia's cyber warfare campaigns. This study provides an understanding of the importance of state cyberspace operations; why the activities are so useful, and how they influence operations that connect conventional and digital efforts.

The issue of separating seemingly isolated incidents from strategic analysis is striking. Until recently, states treated cybersecurity incidents as isolated occurrences, without analyzing them through the lens of strategic significance; this treatment was because they viewed cybersecurity mainly as a technical matter. That is not the case, and never was.

Cybersecurity is not just about the protection of infrastructure used to transmit, receive, store, process and analyze information. It is also about the content, the interaction between the information sphere and the information resource – in this case, the individual. The population, the users, create vast amounts of data. Cybersecurity incidents have an impact on the national security, economic stability, and well-being of the population.

Cybersecurity was for a long time seen as the domain of highly-skilled technical people located in a basement providing IT maintenance. Considering cybersecurity a technical matter has been a luxury that has led us to the unintended consequences of underestimating its importance for any process or entity today.

The continuing problem is that individuals in decision-making positions within the government rarely appreciate the role of information security, cybersecurity and IT services. For too long, too many have seen it as simply the person changing printer toner or maintaining the network, or the one responsible for securing the flow of information in the institution. From a psychological perspective, abstract concepts such as cyber, IT, and InfoSec are hard to visualize. As such, they are harder to understand as challenges and perceive as threats, and more difficult to communicate to the right people in the right positions.

Nevertheless, many challenges and threats abound in the areas of cybersecurity, cyberspace, the information sphere, and the single global information space. Law enforcement agencies emphasize various aspects of cybersecurity, such as analysis of technical knowledge or criminal behavior. Others analyze national security matters affected by recent technological advancements. A strategic mindset requires mastery of both the policy and technical worlds, to be able to sift through enormous amounts of technical information and then correlate it with real-world events. The strategic mindset is centered on the confluence of international politics, national security, one's own interests and those of one's strategic adversaries.

Cybersecurity challenges and threats are not only about the illiteracy of policymakers regarding the technical aspects of informatization. The lack of understanding on the technical level about the potential consequences of seemingly unrelated incidents is the complementary problem. The technological world lacks the in-depth analysis of an adversary's objectives or a correlation of incidents with events in the physical or international security domains. Knowing an adversary's intent and understanding what informs the strategic mindset of the opponent – for example, his use of doctrine – helps one to contextualize what is happening on a much broader scale.

This study examines information warfare, not cyberwarfare. The Russian Federation, the subject of this study, engages in what it considers information – not cyber – warfare. Its efforts include cyber but encompass much more than cyber, hence *information warfare.*

Cyberspace, as part of information warfare, has become a decisive arena. It opens up new dimensions of conflict with an inbuilt psychological impact. By using information warfare methods to attack the centers of gravity and critical vulnerabilities of adversaries, it is possible to win militarily as well as politically against an opponent, at a low cost, without necessarily occupying an enemy's territory. Using information methods delivers productive results vis-à-vis societies heavily dependent on information resources, computational power, and digitalized processes.

Information warfare is as a potent tool for power projection either in a stand-alone mode or conjunction with military operations. Like other major countries, Russia is developing capabilities for *information warfare* (IW) and *information operations* (IO). IW and IO are much older than the concept of cyberspace as the West perceives it. However, cyberspace is the perfect ground for exercising IW and IO. Within the Russian executive and military branches several organizations are responsible for handling information warfare, psychological operations, and deception campaigns, or *maskirovka*, dubbed "content-based operations." Another example is the "code-based operations" concept based on programming and impact from malicious code that exploits the technical side of the information sphere.

The reappearance of old concepts tied with new weaponry emanating from digitalization and new technologies will significantly impact not only on the way strategic objectives are attained, but also the way one defines the strategic goals themselves. In the past, nations achieved strategic objectives through securing military victories over the opponent, usually finalized either with occupation of territory, destruction of the opponent's infrastructure, or elimination of his command-and-control structure; that is, by subjugating through military means. Today, strategic objectives could be attained through *influence campaigns* against the leadership and population to undercut self-preserving mechanisms and behavior. Studies of why deception operations using cyberspace are so efficient against political systems and societies are questions for the realms of psychology, sociology and political science. However, to be able to understand an activity's efficiency (and what is happening), it is essential to grasp why the events happened, and how they occurred in the sense of propagation, techniques, and tools. One can divide this into two subgroups: impact on society and individuals on one side, and the historical-technical understanding of the activities on the other. This study provides a perspective from the doctrinal/historical toolbox and the repurposed analog concepts for the digital era.

ACKNOWLEDGEMENTS

I would like to thank the George C. Marshall Center for selecting me as their Alumni Scholar for 2018. Without their support and benevolence, I would not have been able to finish this. Professor Phil Lark and Maj. Istvan Feher were instrumental in polishing and cleaning up the linguistic mess I made. Thanks go to my team for covering for me while I was out of office doing research and writing. To my friends, who encouraged me and asked me inquisitive questions such as — "Do you really want to write a book? or "Let's go to a bar instead!" — thank you for all the support. Big thanks go to my colleagues, Martina and Peter, for reading parts of the book and being eager to read more. And of course to Sean Costigan, the puppeteer behind the scenes, who transformed this study into a book.

Last but not least: thanks Mom for bringing me up the way you did – that I am able to materialize seemingly intangible ideas into something I hope is helpful to my peers in the cybersecurity and national security domain.

INFLUENCE CAMPAIGNS

Influence campaigns surround us every day, whether with malicious intent or for product and service advertisement. Asking whether influence operations are on the rise is irrelevant. The more pertinent question is instead whether one is susceptible to them and, if so, how they work so one can understand them and become resilient against any adverse effects they might present.

Public debate about these activities has gained significant momentum in the past few years. However, we lack a deeper understanding of the phenomenon. When it comes to influence-operations aimed at changing our threat perception or electoral behavior, a simple declaration does not help us in understanding how to become less vulnerable. There are numerous articles and studies with the quality of descriptive campaign analysis conducted via non-governmental organizations, media outlets, think tanks, and directly via government efforts. The question of why these activities are productive and what tools are employed is central to safeguarding against further exploitation, but one must first identify these malicious efforts and comprehend the goals of the adversary.

While looking at the typology of tools and their effectiveness within influence campaigns, it is essential to recognize that influence campaigns seek to achieve a tactical, operational or strategic aim. When considering influence operations related to military objectives or political struggles, one sees various tools being applied to pursue the goals of such operations. The conduct of influence campaigns is usually part of a broader set of activities, such as those visible in so-called grey-zone conflicts in recent years. As Mazzar defines grey-zone conflict attributes,[1] they:

- Pursue political objectives through cohesive, integrated campaigns;
- Employ mostly non-military or non-kinetic tools;
- Strive to remain under key escalatory or red line thresholds to avoid outright, conventional conflict; and,
- Move gradually toward its objectives rather than seeking conclusive results in a specific period of time

A fundamental way to understand influence campaigns is by categorizing their events and influence activities and then, where possible, correlating the findings with known patterns from the last century. While cyberspace has introduced new avenues of information dissemination, the majority of influence operations have targeted well-established societal functions such as commercial advertising, elections, political contests, democratic foundations of political systems, and individual perceptions, to name a few. These functions have been present in the past and have been targeted by tools exclusive for their time without a cyber-domain presence. Influence activities include:

> ...state control over media, subjugation of political parties, fraudulent elections, using blogs and trolls and social networks for information operations, pacification of in-

tellectual opponents, such as non-governmental organizations, attacks against proclaimed enemies of the state - political opponents domestically and internationally, creation of pro-government mass media engaged in disinformation campaigns, dissemination of disinformation to undermine conventional press, gaining political influence in foreign countries, production of pro-government non-governmental organizational advancing the foreign policy goals favored by the establishment.

If an exemplary list is divided by aim and action one can readily associate it with a potential cyber-component to achieve the desired objective. For illustration, the objectives are labeled *aim* followed by the proposed activities to achieve the objectives. The chart is complemented with the cyber component classified either as code based action or content based.

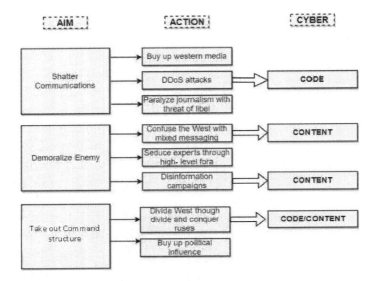

Figure 1

WHY IS ANALOG HISTORY IMPORTANT TO THE DIGITAL PRESENT?

Cyberspace has changed neither the strategic intent, nor the perception of adversaries and threats. It has provided new opportunities, but also vulnerabilities and ways opposing entities can influence outcomes. On a less positive note, it has enabled one's nation to conduct influence operations more efficiently and to exploit its advantages when on the offensive.

However, there is a generational knowledge gap. Cybersecurity technical analysts focus mainly on the technical and scientific aspects of cybersecurity incidents, cyber attacks and computer network operations. Seasoned analysts from an older generation must contextualize seemingly unrelated dots of cybersecurity incidents. Contextualization requires carefully working with technical staff. Unfortunately, fusing the knowledge base between these two types of analysts has been strikingly absent from strategic analysis. One reason why that may be is that the U.S. and European approaches to national security lack an understanding of where the threats reside. Analysts and policymakers operating at the strategic and operational levels are not adequately acquainted with the technical background necessary to grasp the threats and challenges coming from cyberspace. Incident handlers and cyber threat intelligence analysts working in cyberspace on a daily basis lack the historical context for how Moscow has adapted influence operations concepts to cyberspace. One must, therefore, study analog history and draw lessons from it, correlate those lessons with cyberspace campaigns, and realize how the two worlds are intertwined.

To summarize, it is of vital importance to study the tradecraft of past influence operations. The study provides insight into the strategic goals and mindset of an adversary, and how these goals affect their cyber tools of choice. One must appreciate the cyber link to analog concepts of *information operations* and their adaptation from past techniques. One must accept that these old concepts rule the "new" decisive arena for future conflicts. Fundamentally, the ideological positions and worldview of the adversary defines the tradecraft, and the adapted concepts predetermine the scope of activities an adversary will conduct to protect their strategic interests. Finally, one must comprehend that contextualizing cybersecurity incidents with physical world events and national security threats is essential for neutralizing or defeating an ongoing or future code-based or content-based operation.

GENERAL INTRODUCTION TO DECEPTION[2]

Deception maintains a long tradition as a tool in military and political conflict. It does not remain unique to the activities of humanity, as examples of camouflage, concealment, and misleading features and behavior are common in nature. Deceptive conduct, camouflage, protective elements of coloration, shape, and masking are present in the realm of fauna and flora. The effectiveness of deception is unquestionable.[3]

> The traits of deception are poetically described by Sir Francis Bacon in his essay "Of Simulation and Dissimulation" from 1625[4]

There are a few basic principles of deception to consider here. If the provided information is false, it is called fabrication. Such disinformation is intended to mislead an opponent. Manipulation is the use of truthful information, but distorting its meaning and purposes by presenting it out of context to imply a false sense of it. Political deception enjoys a close relationship with military operations and is considered the easiest to conduct.[5] Hiding one's strategic objective from one's adversary is the most common form of political deception.

Military deception is more clandestine. Military deception can be subdivided into general categories, such as passive and active deception. Passive deception hides intentions, military readiness and capabilities from an opposing force.

The definition of passive deception encompasses hiding something that already exists. Active deception on the contrary, creates false assumptions; for instance, of something that is nonexistent. This might include the projected state of readiness, military might, or intent. Another category in understanding deception is the level of specificity of the ruse.[6]

Donald Daniel and Katherine Helbig divide deception into A type and M type classes.[7] The A type stands for increasing ambiguity to promote and disseminate confusion that leads to distraction. This A-type deception is often used today because communication channels allow widespread distribution to sow confusion. The M type, based on the misleading-variety principle, creates false assumptions based on provided information and leads an adversary to believe a non-existing model of reality.

> As General Sherman put it, "the trick is to place the victim (of the deception) on the horns of a dilemma and then to impale him on the one of your choosing."[8]

Two other well-used methods of deception are *cover* and *conditioning*. Cover disguises an incoming hostile activity by filtering it through a non-threatening activity. Those can be, for example, military exercises, or, in the diplomatic world, fake negotiations. Conditioning is a method broadly used in the military, politics – but also in advertising. The conditioning works by repetition; in the military domain, for preparations to attack that in the end do not happen. Repeating a position or negotiating a position that is not seen through and carried out is an example of conditioning in politics. The preparation for a hostile act, and then repeatedly not carrying it out, leads an adversary to a false sense of security. This false sense of security is a concept presented by Caddell – familiarity breeds contempt.[9] One of this method's objectives is to desensitize an opponent to a looming threat.

> In 1947, General Eisenhower, as Chief of Staff of the Army, and reflecting on his wartime experience, wrote a memo to his director of Plans and Operations saying, "No major operations should be undertaken without planning and executing deception measures."[10]

According to the *Deception 101 - Primer on Deception*, the U.S. view of strategic, operational and tactical deceptions are:[11]

- Strategic deception intends to "disguise basic objectives, intentions, strategies, and capabilities."
- Operational deception, which confuses an adversary regarding "a specific operation or action you are preparing to conduct."
- Tactical deception is intended to mislead "others while they are actively involved in competition with you, your interests, or your forces."

The categorization of deception is often not tied to the method or means of deception, but to the objective of the trick. A nation can use a very primitive type of trickery to achieve a strategic goal or a well-crafted and complex deception campaign to gain a tactical result. In conducting a deception campaign, it is best to remember that ignorance, arrogance, and fear all complicate one's ability to detect false information.[12]

Prejudice and predetermined thinking cause cognitive dissonance. Cognitive dissonance is an essential part of the deception process, especially among the general population and political leadership. Ignoring vital information because it does not fit the preconceived image of the outside world and its opinions and theories leads to cognitive dissonance. Caddell also stresses the importance of inertia of rest. Strictly speaking, this is the inability to adjust to reality when facts have refuted assumptions. Caddell writes: "This refers to a tendency of people to believe certain assumptions remain valid even after they have been undermined by events. In physics, 'inertia of rest' refers to the tendency of an object at rest to remain at rest until acted upon by an outside force."[13]

To summarize, the core of a successful deception operation lies in exploiting the adversary's sensory awareness and cognitive functions altered by influence. Caddell introduced two terms in his primer. Based on the Soviet school of deception, one recognizes disinformation and maskirovka from the military perspective – "Dezinformatsia: The dissemination of false or misleading information intended to confuse, discredit or embarrass the enemy."[14] This definition comes from the works of Marshals of the Soviet Union, Grechko and Ogarkov[15] –

> Maskirovka: A means of securing the combat operations and daily activity of forces; a complex of measures designed to mislead the enemy as to the presence and disposition of forces and various military objects, their condition, combat readiness and operations and also the plans of the commander. Maskirovka contributes to the achievement of surprise for the actions of forces, the preservation of combat readiness and the increased survivability of objects.

CYBER POWER IN RUSSIA

> Invasion of ICT calls for information superiority, and long tradition of deception campaigns provided for experience.

The perception of cyber power differs significantly between the Russian approach and the U.S. approach.[16] The view shared among Russian academics and military experts is that cyber warfare encompasses not only information-technological but also information-psychological aspects.[17] The understanding of Russian academics and military experts is that Russia is experiencing an ongoing attack in a hostile cyber environment through content aimed at undermining Russian strategic interests. Grasping the difference in the understanding of cyber power between the U.S. and Europe and the Russian Federation is essential for classifying the activities conducted in cyberspace. The differences are based on the understanding of cyber power itself, which eventually demonstrates that the opposing parties employ different toolboxes and sets of activities. Russia more broadly understands cyber power and information warfare as a holistic concept.[18]

Summarizations have been made to the point that information war – "informatsionnaya vojna" – includes computer network operations, electronic warfare, psychological operations and information operations.[19] There is a division between hostile code and hostile content,[20] two distinctive measures not often regarded in the West as a part of one campaign. Information-technical operations serve as the distributor of hostile code and hostile content. Those range from influence operations to physical damage to infrastructure, as well as information-psychological operations to attack the perceptions and morale of entities or entire societal groups as described by Major-General Sheremet.[21]

The intent is to achieve digital disruption, as well as psychological subversion. Beyond disrupting the ability of communication, for example, another critical component to a successful information warfare campaign is to disorient, delegitimize and discredit the adversary and demoralize the will of the opponent using hostile content and hostile code.

There are three sources of Russia's holistic approach to information warfare. The first is the military-technical revolution, an information technology revolution in military affairs. During the Cold War bipolar era, both superpowers faced challenges regarding the incorporation of information technology. The concept, however, is still driving military innovations as the technology evolves. It is safe to state that the militaries in the present, multipolar world, are not spared the challenges of incorporating new technologies and concepts. We are far from the final stages of adapting technological advancements into our militaries. The task is almost the same as in the arms race: to adapt and overcome the opponent by embracing new technology or adapting old concepts and tools to new realities. In cyberspace, we are witnessing both incorporation of the new and adaptation of the old, especially in the realm of information operations; refitting analog concepts to serve in the digital era is more than a choice – it seems to be a necessity.

Mary C. Fitzgerald correctly pointed out in 1997:

> According to the Russian military, superiority, in the new Revolution in Military Affairs (RMA) proceeds from superiority in C4ISR systems: 1) reconnaissance, surveillance, and target acquisition (RSTA) systems, and 2) "intelligent" command-and-control systems. Information technologies are now said to be "the most formidable weapons of the 21st century" – and comparable in effects to weapons of mass destruction. Indeed, they constitute the essence of the new, 4th RMA. The Russian politico-military leadership therefore is engineering a dramatic shift away from Industrial Age material-intensive systems and toward *Information* Age systems: away from ballistic missiles, submarines, heavy bombers, tanks, and artillery and toward advanced C4ISR and EW systems. Warfare has indeed shifted from being a duel of strike systems to being a duel of information systems.[22]

In the past, militaries of the industrial era were constituted of mass-conscriptions. Also the concept of "total war" as coined by Erich Ludendorff,[23] was the framework of war efforts. That meant everything in the state was subordinated to the war efforts. Today, as we are entirely reliant on modern technologies, the whole command and control process – from acquisition of intelligence, to analysis, through the decision-making chain is based upon the undisrupted flow of information. We are witnessing so-called reconnaissance strike complexes. The technology, which provides us with the ability to gather and analyze vast amounts of data and communicate at previously unprecedented speed, also created an inherent vulnerability in the decision-making system.

> Warfare has indeed shifted from being a duel of strike systems to being a duel of information systems.[24]

Today, you can defeat the reconnaissance strike complex via an information strike. This strike can be very simplistic. Thanks to technology and the channels to disseminate information, there are ways to obscure true intentions, plausible deniability, delivery platforms and dissemination speed – you can influence whole societies.

The information strike, coined as information warfare by Andrew Marshall of the Office of Net Assessment,[25] heavily depends on the shift from industrial age militaries, consisting of large-scale numbers and destruction of the opponent, to the growing dependence on information and cognitive perception in decision-making. This approach, the delivery of falsely crafted intelligence or tampering with the decision maker's sensory awareness of the battlefield, is nothing new. It has been part of military strategy and operational art as long as humans have fought against each other. The new factor is the technological underpinning of today's information warfare activities, which provides the opponent with efficiency and tools technically incomprehensible before the 1990s.

The second source is the concept of active measures. This term encompasses several techniques described in "Soviet Influence Activities, A Report on Active Measures and Propaganda, 1987–1988." Department of State Report (1989):[26]

Disinformation and Forgeries

Disinformation, a deliberate attempt to deceive public or governmental opinion, can be oral and/or written.[27] Forged documents are frequently used in attempts to discredit individuals, institutions, or policies in such a way to damage U.S. foreign policy interests.[28]

Front Groups and Friendship Societies

Fronts normally present themselves as non-governmental, non-political organizations engaged in promoting desirable goals such as world peace.[29]

Non-ruling Communist and Leftist Parties

Contacts with these parties are usually overt and often are used to persuade the parties to carry out specific political action or propaganda campaigns on behalf of the USSR.[30]

Political Influence Operations

Agents of influence disguise their KGB connection while taking an active role in their nation's governmental, political, press, business, labor, or academic affairs. Their objective is to convert their influence in those realms into real policy gains for the Soviet Union. At times, the Soviets use unwitting contacts to achieve similar results.

Active measures were adapted to serve its operational and strategic purpose in the digital era, underpinned with the options available through modern technology. Active measures encompass a wide range of activities, but to understand it in the original context the term is *maskirovka* as to "mask something." The word maskirovka has no direct translation in English, but it covers activities such as the use of dummies, decoys, execution of demonstration maneuvers, camouflage, concealment, denial, deception, and disinformation. Maskirovka is by no means a new concept original to cyberspace. On the contrary, it has been part of the operational art and intelligence tradecraft for centuries – systematically taught at a military school founded by the Czar Nicholas II. For Russians, maskirovka is a crucial component of information warfare, primarily due to the level of penetration of technology into military activities. [31]

The cyber domain, with its attribution problems and fuzzy boundaries, is the perfect environment for conducting information operations. Because of its specific nature, cyberspace is considered a force multiplier. In the strategic context, it provides smaller and mid-size entities, states and non-state actors with a comparatively increased advantage previously unheard of. The distance, commitment of forces, risk of exposure and attribution, level the playing field for all actors in cyberspace with no regard of size, geographical location or conventional military maturity.

For any strategist or military planner, the ability to operate rapidly against distant forces of the adversary without the commitment of combat personnel and the ability to act in secrecy by minimizing exposure leading to attribution and risk of counterattacks is like a dream. Also, the ability to conduct information-technical and infor-mation-psychological operations as means of disruption, denial, destruction or subversion of critical national infrastructure without necessarily reaching the threshold of armed conflict constitutes near-limitless possibilities. Active measures, which were confined to the physical domain in the

Figure 2 Czar Nicholas in field uniform².

analog era, were a natural choice of a "tool-box adaptation" to be utilized in cyberspace.

The expansion in cyberspace has significantly amplified active measures. The strategic maskirovka concept is very straightforward. It aims at manipulating the decision-making process of your adversary to maneuver his strategic behavior in the desired direction. This concept is one of the critical sources of Russian cognitive-psychological inspiration when conducting cyberspace related activities.

The third source is cybernetics, a subfield of applied mathematics. Cybernetics, distinguished from "cyber," is a discipline, which connects the realms of exact mathematics and social sciences. Cybernetics explores the laws or the nature of decision-making processes in complex systems or systems of systems. This hints to the fact that cybernetics, in Slavic languages "kybernetika," consists of technological/digital and cognitive perceptional components, which again is of great importance in today's digital age, where we consume content provided by technical means.

These three sources of Russian operational art in cyberspace collude into information struggle, which is much broader than the U.S or European-centric perception of cyber warfare or cyber operations.

To portray how deeply the concept of maskirovka runs in Russian strategic thought and how vast its historical roots are, we need to dive into history. The depicted Czar Nicholas II in an army uniform has more to do with information operations in cyberspace than you might anticipate. Founded in 1904, during the reign of the Czar, the Higher School of Maskirovka provided for the frameworks used today.[33]

The foundations of the Russian approach to information operations go even deeper into the late 1860s, when the Czarist Secret Service Okhrana was operating with deception against anarchist entities.[34] The school provided the basis for maskirovka concepts and created manuals for future generations of intelligence officers in the Soviet era, manuals upon which we draw lessons for today in the information struggle with the Russian Federation.

Apart from the somewhat limited scope of information operations by the U.S. and European allies (limited by the condition whether you are at war/armed conflict or conducting operations in peacetime by intelligence entities), Russian strategic thinking recognizes information threats to be permanent, consisting of hostile code and hostile content. To them, the danger is imminent, and the doctrines emphasize the regular role of information operations in peacetime as well as during hostilities.[35]

The differences are also to the extent and spheres of activities conducted. Russian military strategists navigate through three different areas of activities. First, the information sphere where the perception of an individual or an entity is formulated, be it cyberspace or a closed environment. The domain is where we navigate to gather information and interact with other individuals or entities. Second, the information itself shapes the perception we create about our environment and circumstances. This is the content that carries the message or meaning to our sensory awareness. The third area of activity is within the information infrastructure. This infrastructure is the technical component, which provides digital manifestation to the first two spheres. The disparity of understanding between the U.S and European-centric perspective and the Russian is because we are aiming predominantly at the protection of the infrastructure, but omit the information sphere and information itself, at least during peacetime. Where "our" cyber operations contain almost exclusively activities aimed at infrastructure, the Russian operational art wholly intertwines the spheres mentioned above, and it is unthinkable to deal with each field separately.

Also, the term "cyber" seems to be used in Russian literature mainly as a description of U.S. or Chinese activities. The Russian perception includes cyber implicitly with electronic warfare, psychological operations, strategic communications, and influence.[36]

In summary, the goal of the information campaign is to manipulate decision-making processes in order to interfere and to force the other side to act upon a picture of reality that the initiator of the campaign produces. It is highly unlikely to achieve a desired result with attacks on infrastructure only. However, by attacking an opponent within the information sphere – and attacking the information itself – obtaining the goal (the idea of producing a false picture of reality without the other side being aware of the falsification) is a probable outcome. This could lead to a desired state of information superiority. Russian strategists strive for disorganizing the enemy as well as achieving information superiority; in fact, they believe the former produces the latter. Thus, to achieve information superiority in cyberspace, they have to disorganize the enemy. Tools used for achieving deception of the opponent range from the maskirovka toolbox to the concept of reflexive control.

REFLEXIVE CONTROL:
RAISON D'ETRE

In 1950s and 1960s the U.S. surpassed the Soviet Union in every economic indicator.[37] Because the two blocks also competed for global political dominance through military might, conventional and nuclear arsenals played a role of utmost importance.

The Soviet Union eventually realized that it was struggling to match technological developments. One such indicator of this Soviet technological deficit was the creation of Directorate T within the KGB,[38] where a particular branch was developed to covertly acquire or steal western technology. The industrial complex in Soviet Russia was struggling to match the requirements for keeping up with the West not only in the arms race, but also in the civilian industrial complex.

Soviet awareness of the pace of U.S. technological development and military superiority[39] in advanced technological conventional warfare induced exploration and research of alternatives to hard power. In the late 1950s, physical and social regulatory systems caught the interest of Soviet scientists. During the research on physical and regulatory systems, the knowledge of maskirovka, or tradecraft in general, inadvertently built upon the ideological predispositions[40] and led to forming a theory that in essence was mathematically based, but also had a stronghold in social sciences, psychology and communication theories. Well beyond psychological warfare, information warfare and information operations, reflexive control was almost a unifying concept regarding how to employ above-mentioned concepts, well beyond regarding their direct or indirect effects on the strategic, operational and tactical level.

The research on reflexive control was tied to prior study of military cybernetics at the First Computer Center of the Soviet Ministry of Defense (military unit 01168).[41] The research objective was the development of methods for the optimization of military decision-making, based on computerization and digitalization.

However, reflexive control is not a panacea. Conducting reflexive control requires hard power capability, disinformation, manipulation, and tools with which to influence decision-making algorithms. Applying reflexive control on a tactical level, the adversary uses information about the area of conflict, about troops, combat status and availability, to make decisions regarding how to proceed and what plan to execute. By correctly employing reflexive control, you take the decision-making mechanism out and control the outcome by controlling the opponent's actions without his knowledge through a previously devised set of steps the opponent follows, in a reality he is unable to recognize as divergent from the facts on the ground.

Reflexive control, as adopted by present-day Russia, gives a reasonably competitive edge and constitutes a vital component of its modern warfare strategy.[42] One of the reasons reflexive control is so effective in the West is that we are too focused on flawless processes and technological advancements at the expense of a vital component: the decision maker. If the decision maker is open to influence, all the processes and technological advancements are in vain. The goal is to flatten technological disparity by leveraging the fact that the struggle is not between conventional firepower or nuclear arsenals; it is between decision-making complexes and the ability and will to deploy the assets at one's disposal. Take away the ability to decide, utilize, alter political will or the ability to execute the commands, and the military arsenal is useless. Therefore, Soviet strategists have studied the importance of indirect approaches to offsetting U.S. superiority.

A THEORETICAL MODEL OF REFLEXIVE CONTROL

Reflexive control is about modeling decision-making systems, understanding them and eventually disrupting them. The main aim is to influence an adversary into making decisions favorable to you as the deceiver.

Decision-making and command and control in the military during an operation is based on information about troops, the ability to fight, resources and merely a "cost-benefit" analysis also framed by the rules of engagement and international humanitarian law. However, if you influence the channels of information and send messages that shift the flow and content itself in a way favorable to you, the opponent will conduct his activities in a way beneficial to you without realizing it.

Conflict here is not between military forces, or political might and international negotiations. It is between decision-making processes. Also, limitations of mental capacity play a role. The complexity of today's world is far reaching beyond the ability to process all the ready information and properly contextualize it. The individual – the decision maker – instead creates a simplified model of reality and then uses this model. The individual then reacts to the limitations of this model.[43]

> According to the creator of reflexive control, rather than looking at conflict as an interaction between two military forces, conflict should be considered as being between the decision-making processes of the two opposing entities.[44]

The inventor of reflexive control, Vladimir Lefebvre, created a modeling system comprising of three subsystems:[45] A model to simulate your decisions, another for your adversary's decisions, and one to make real decisions.[46] Lefebvre determined that if the Soviet Union could get into the decision-making process and understand the procedures and ways of how decisions are made and influenced naturally, the Soviet Union could provide the adversary with information and conditions which might lead the adversary into taking predetermined choices in favor of the USSR.[47] Imagine the potential of this theory in influencing the decision-making algorithm.

In the past, conventional means such as espionage, disinformation, NGOs and buying up political influence were the tools of deception. The theory never really reached its full potential, due to limited distribution methods and limited ways to deliver the particular vehicle of maskirovka to the masses. Today we have cyberspace, with its inherent ability to communicate instantly, reach societal groups directly and exploit grievances to tailor the activities carefully – and with cyberspace serving as an information highway. It is a gift for those willing to employ maskirovka principles, and hell for the unprepared. Shifting perceptions not only of decision makers, but of society, was never more easy. Reflexive control is all about conveying pre-crafted information to incline the opponent to voluntarily make the predetermined decision you desire.[48] This is an old concept that also worked for the Allies in World War II.[49] However, the difference is not only in the wartime/peacetime employment but also in the toolbox and activities used. Planting information to create a ruse is a difficult task. However, reflexive control is intended for strategic, operational and tactical purposes, employing civilian and military assets in a continuous campaign sustained over an extended period to change the behavior in the decision-making system in your favor. It is not about planting information to achieve a military objective and divert the focus of the opponent. Reflexive control serves to undermine the very decision-making system itself, to make it favorable to the projector and thus to project power without committing significant military or political resources, nor meeting the acknowledged threshold of meddling in a sovereign's internal affairs.

Reid in his *Reflexive Control in Military Planning*[50] laid out elements important for creation of perception necessary in military decision-making. Those elements are:

1. The size and characteristics of one's own forces
2. The size and characteristics of the opponent's forces

3. The physical environment within which conflict occurs
4. The history of actions by the two sides
5. The current evolution of events
6. The objectives and constraints of the opponent.[51]

The first three elements are essential in the scope of situational awareness. The distortion of the perception of reality is done by attacking the sensory sentience. The second three elements are based on the analysis and knowledge of the opposing entity.

In general, one of the ways to exercise reflexive control is to manipulate sensory awareness: how you process information, leading to a different perception of reality. Also, it is important to hide true intentions from the opponent, so that the real strategic aim cannot be revealed, analyzed and thwarted.

There are four general prerequisites for conducting reflexive control:

1. Manipulation of sensory awareness

Reflexive control as a product of the Marx-Leninist paradigm.[52,53] The decision maker's cognition – perception of the outside world – is based on reflection of the material world. The ability to make decisions lies in the ability of intelligence to analyze the outside world. However, intelligence is dependent on sensory awareness of the world. Therefore, the content and dimensions of consciousness are determined by what we see and hear, which defines our reactions and actions.[54]

2. **Hiding true intentions to the opponent**

Keeping the opponent in dark regarding the actual plans leads to easily manipulating his sensory awareness – thus tighter control of his decision-making process. The human mind processes everything available to the task it is focused on. To keep the decision-making algorithm unpolluted, it is vital to hide the real intentions. The decision maker has to have the feeling that he made the decision independently, based on all available information, otherwise he might not pursue the subsequent activity desired.

3. **Influence the opponent's information resources**[55]

Information resources of the entity, or decision-making complex of the entity, are essential for the functioning of the object and for command and control in general. These are the sensors, channels, and memory necessary for utilizing information for the benefit of decision-making. They serve as communication tools, to store information and for the management of information.

The list of information resources is described well by Timothy Thomas:

- Information and transmitters of information, to include the method or technology of obtaining, conveying, gathering, accumulating, processing, storing and exploiting that information[56]

- Infrastructure, including information centers, means of automating information processes, switchboard communications and data transfer networks.[57]
- Programming and mathematical means for managing information.[58]
- Administrative and organizational bodies that manage information processes, scientific personnel itself, creators of data and knowledge bases, as well as personnel who service the means of *informatizatsiaya* (informatization).[59, 60]

If we look closely at the definition of information resources, it is imperative to understand that these information resources were defined in the age of conceptualizing reflexive control – in a world where information was still stored on paper, by analog means. The modernization of communication was in its birth stages, and nobody could foresee the complete dependence on cyberspace and information technology we are witnessing today. The information resources, existing regulatory systems and decision-making complexes were defined and targeted based on the technology present at that time. However, with the influx and almost revolutionary pace of digitalization, these concepts of information resources gained in significance and became vital to our societies, governance and command and control. The activities of maskirovka utilize hostile code and hostile content. Proper utilization of hostile code and hostile content activities, orchestrated within the reflexive control framework would have a debilitating effect on the opponent. It would lead to the inability of the opponent to respond to hostile activities in general. This would be achieved due tom refocused priorities and outcomes of decision-making processes.

4. Combination of tampering with filters (data processors) and sensory awareness (images of the outside world)

Models of communication best describe the means of tampering with filters and sensory awareness. To understand reflexive control, we have to look into ways we transmit and process messages and information.

Even the very basic "message influence" model provides us with at least four entries by which we could tamper with the sensory awareness and the data processor itself (either a decision maker or computer).

THE MESSAGE INFLUENCE MODEL

Figure 3

This model is "sender oriented," with the assumption that if you send the message correctly, it will get through to the intended receiver/audience. There are no means of influence from the outside world in this model. However, numbers were added to the model to show tools of reflexive control, when used adequately, suggest potential vectors through which the audience's perception of the message source may be tampered with. Thus, the message may be altered or replaced by an entirely different one for distribution. Further, the channels can be tampered with, either by disrupting the signals through efficient compromise of the channel, or by creating different channels to replace legitimate ones. Finally, by individually crafting information/messages using emotional, linguistic and psychological means, you can alter the perception of the audience by merely having them respond more emotionally or less, depending on your preferences. Specific societal groups are sensitive to tailored forms of information such as graphic imagery and not just strictly content, which can induce emotional distress, nurturing confirmation and cognitive biases.

Perceptual biases are developed on individual and societal perceptions of the environment and the process of perception itself. Biases limit the accuracy of perceptions. Mental or emotional factors do not constitute a base for cognitive biases; they arise from the way mind works.[61] Biases are based on the fact that in our minds, we construct reality instead of recording it. This is dependent on the so-called sensory awareness that is the primal data input in the construction of our perceptions of reality. (For more on biases and their importance for deception and intelligence, the work of Richards J. Heuer, Jr, *Cognitive Factors in Deception and Counter deception,* is a good primer.)

It is a false assumption that if we send the message to the audience the proper way, it will be received unbiased, with its intended meaning. This was factored in when crafting disinformation in the past. According to a study titled "Strategic Communication on a Rugged Landscape,"[62] the message sent very often was not the message received, its reception and understanding being heavily contextualized by background attitudes and perceptions. Taking the basic "sender-oriented" model and placing it in the environment of hostile code and hostile content provides us with a simplified perception of the many entries where the information transmission between sender and audience is distorted. Distortion is achieved by adding the basic characteristics of information processing such as perception bias, acceptance, modification or refusal.

The author of reflexive control argued, "In making his decisions, the adversary uses information about the area of conflict, about his troops and ours, about their ability to fight, etc."[63] He argued that it is possible to influence the channels of information and send messages that shift the flow of information in a way desirable to the opponent, making the opponent control the situation.

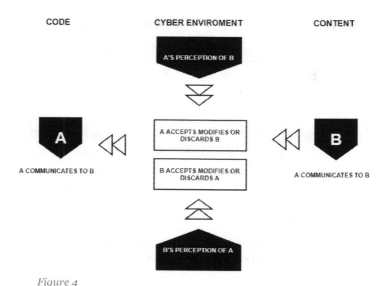

Figure 4

What better environment to do this on the scale previously not imagined than in cyberspace, employing reflexive control on a massive societal, political and military level? The decision-making process relies on available information. To make better decisions is one of the reasons humankind collects information. That is why intelligence agencies strive to collect and analyze information that is not freely available. Vast sums and effort were, and are, invested throughout human history in acquiring information in order to improve decisions. Distorting the ability to seamlessly navigate through data and information, the ability to transmit without corruption, and to accurately analyze and understand the real world, leads to flawed decision-making and crippled command and control.

The Soviet military school of thought offers three distinct ways of influencing the objectives of the opponent:

1. Show of force that leads to the adversary's perception of unattainable objectives.
2. Provide the adversary with such amount of ambiguity and uncertainty about your goals, that the adversary is unable to define any aim that has an acceptable result for all possible sets of events.
3. Exhibit a threat of such a magnitude that countering the threat governs the objectives of the opponent.

A deep understanding of an adversary provides successful reflexive control. It is not only the cognitive predispositions that are important, but also the analytical processes, a perception of the world and concepts helping to understand the reality. These elements which define behavior are called filters. A filter is a sum of inputs, called a set.[64] The set defines the enemy – how he behaves in certain situations, how he interacts and what his line of thought is when making decisions. The reflexive part is therefore the study of the set or filters and subsequent exploitation; in short, finding the weakest link. With the digitalization of information acquisition, processing and analysis, there are in fact two filters to study and exploit. These filters are the individuals and the computerized information processes.[65] In the conduct of reflexive control, the prime task is to convey motives and reasons to the opponent. By transmitting motives and intentions, the objective system that is under the influence of the controlling organ is taking actions and making decisions independently. The reflex involves analysis and imitation of reasoning by the enemy, and the imitation of possible behavior that leads to unfavorable decision-making as the objective.[66]

Reflexive control in its essence is a long-term influence campaign, aimed at feeding the adversary with information with an intended impact on the moral values, psychological state or even the decision maker's character. The longevity of the campaign is important, as psychological studies show that sustained exposure to information, irrespective whether the information is true or false, tends to create a perception of the information as true.[67] This is true even when the evidence is present to prove the information to be false. Therefore, debunking false information and refuting provided data as incoherent and leading to false assumptions is ineffective in countering disinformation influence campaigns. Also, the more the data seems credible despite being disinformation, and the more it is tailored to seduce the opponent's cognitive biases, the more effective it is.

DECEPTION PREREQUISITES[68]

The deception equation has two main actors, the victim and the deceiver. Apart from choosing the objectives to pursue, the deceiver also has to pick what perceptions will be reinforced, what is to be subdued carefully, which communication channels are to be used, and whether to employ the unwitting agent's methods of deception.

PREREQUISITE A
Set the deception objective

In the civilian world, setting the objective of deception would be used to gain competitive advantage. In the military world, deception is related to survival and strategic surprise.[69] Although very broad in potential, several objectives, which work as prerequisites for deceit but also as objectives themselves, are: Targeting cohesion, spreading distrust and discontent, undercutting national interests, enhancing lack of trust towards institutions and authorities, attacks on liberal order, undermining the electoral processes, crippling command and control, weakening the entity in the international community, and acting against self-interest.

PREREQUISITE B
Deception planning

For any deception to be effective, rigorous planning is a must. The deception aims to incite preplanned behavior in the adversary: the perceptions to be exploited, cognitive biases, sensors to be tricked, and channels to be influenced. All this has its place in deception planning to achieve the desired behavior. Defined by Panagiotis:[70]

> In an information framework, identifying the actions required by the target, determining the perceptions that will induce the target to take these actions and developing the deception story that will lead the target to these perceptions represent the transformation of knowledge into information that will be presented to the target's military and intelligence analysts as well as decision makers.

PREREQUISITE C
Methods of deception

The third step is the methods of deception. Employment of only one method rarely achieves deception. The overlap between methods necessitates rigorous planning. Active deception requires decoys, active spreading of information and imitating or misrepresenting. Passive deceptions are about masking, hiding, and denial. Methods are often limited or defined by channels available or chosen. Some methods are also channel-specific, such as electronic warfare, information operations or computer network operations.

Methods and channels define the parameters of the information to relay to the deceived entity. The channels differ based on the method and intent, but also based on the capacities of the deceiver. Among historically proven channels are politicians, diplomats, intelligence officers, defectors, media, leaflets and modern mass media. The information transmitted to the object of the deception might get corrupted on its way from sender to receiver. The channels are subject to intended or unintended corruption at multiple points, therefore it is imperative for the deceiver to have control over as many channels as possible to ensure the quality and quantity to achieve the deception, and not a failure in conveying distorted information. One way of controlling the channels is to create them. Online media outlets, working as proxies to the deceiver, are ready examples. Having control over the communication networks also ensures a certain degree of supervision over the transfer of the information. With digitalization, new possibilities extend beyond regarding the variety of sources or quantity of channels. Another factor that has an impact on deception methods is the ability to deliver perfectly tailored information to individuals based on their consumer behavior and political perspectives in cyberspace. A further factor is the enhanced application of forgeries. This enhancement is not only the use of images and text processors to falsify maps, official documents, and pictures, but also the emergence of artificial intelligence, allowing for the "deep fake."

PREREQUISITE D
Cohesion in execution
Coherence across strategic, operational and tactical levels is essential. The execution of the deception is at the mercy of the policymakers and planners having the operation or deception

activity under control all the way down to the tactical level of troops in the field.

PREREQUISITE E
Credibility of deception
The information fed to the adversary has to be credible, or at least to have the purpose of achieving the intended aim. The more one knows about the adversary, the better positioned one is to make the deception credible. Even distorted data has to have an underlying motive; otherwise, the data is dismissed, and the deceived becomes suspicious.

PREREQUISITE F
Technology
The technology for deception has more or less been present for hundreds of years as a useful instrument. However, with cyberspace, the information sphere, digitalization, dependence on sensors, computational power, databases to store information, processors to process and analyze information, technology is more important in deception than before. This level of fabrication and modification possibilities provides a break-through in deception options. The introduction of new deception methods, as well as the adjustment and repurposing of the old ones, allows the deceivers to target either individuals or societal groups with surgical precision.

REFLEXIVE CONTROL AND DECEPTION

The previous chapter introduced several prerequisites for conducting deception. This chapter will present examples of reflexive control in use. Military thinkers Tarakanov, Korotchenko,[71] Komov[72] and Ionov[73] provide examples as will the study of reflexive control in military planning by Clifford Reid.[74]

- **Transfer of an image of the situation**
  The transmission provides the enemy with a picture or perception of the deceiver's choosing. It is usually done by distorting the reality, providing false information or attacking sensory awareness. One transfers the image by a variety of methods from merely delivering incomplete information from a trusted source to influencing the acquisition, processing and analysis algorithms.

- **Creation of a goal for the opponent**
  This might be done by influencing the adversary, by providing information that creates a false objective for the deceived. The adversary will then pursue in the deceiver's favor.

- **Transfer of a decision**
  The critical component of this method is trust between the deceiver and the deceived. Previous behavioral patterns and

prior encounters are the keys to conducting this method.

- **Formation of a goal by transferring an image of the situation**
  This is accomplished by providing the adversary with a picture that lures him into pursuing a target based on the flawed view of the reality.

- **Transfer of an image of one's own perception of the situation**
  This transmission is achieved by leaking seemingly legitimate intelligence or information that leads the opponent into a flawed estimate of your perception of reality, ultimately changing his perspective on your impressions of the situation

- **Transfer of an image of one's own goal**
  Providing the opponent with a feint goal, which provokes, for instance, lowering defenses or change in attitude or perception of the situation, or leading to an escalation in the direction unfavorable to the opponent.

- **Transfer of an image of one's own doctrine**
  Projecting false pictures of processes, decision-making patterns and priorities to induce the deceived to try to exploit false weaknesses. This method is closely tied to

military exercises showcasing false procedures and military readiness.

RUSSIAN DOCTRINAL PERSPECTIVES ON CYBER POWER AND INFORMATION THREATS

The official declaration of the role of information operations and the information sphere in the strategic and doctrinal documents gives insight into what Russian leaders believe are their nation's security challenges and priorities. Importantly, it illustrates how the Russian government wishes others to perceive its posture on information security[75] and cyberspace issues.

To tie together the strategic posture of Russia and the position of Russian military thought, below are examples of high ranking military officials speaking of reflexive control and information operations in general.

As noted by Thomas,[76] Major General N.I. Turko, an instructor at the Russian Federation's General Staff Academy views the importance of reflexive control accordingly:

> *Reflexive control is an information weapon that is more important in achieving military objectives than traditional firepower. [and viewing]...reflexive control as a method for achieving geopolitical superiority.*[77]

Having a Major General – who is directly involved in shaping the doctrinal policies and educating future military leaders of the Russian Federation – take this approach underscores the significant role of reflexive control as an information weapon. The admission of the central role of reflexive control in Russian conceptions of information warfare, and reflexive control's potential use against information resources to destabilize the geopolitical balance signifies its importance, not only as an information weapon, but as a central concept of operations in cyberspace as we know it. As stated by General of the Army, Viktor Nikolaevich Samsonov, Chief of the Russian General Staff, on December 23rd, 1996:[78]

> The high effectiveness of information warfare systems in combination with highly accurate weapons and nonmilitary means of influence makes it possible disorganize the system of state administration, hit strategic installations and affect the mentality and moral spirit of the population. In other words the effect of using these means is comparable with the damage resulting from the effect of weapons of mass destruction.

This statement directly supports the highly valued position of information warfare in the strategic military thinking of the top brass of the Russian General Staff. It also relates to the documents presented in this chapter, where information warfare and means of influence pose a threat to the Russian Federation. Russia, wary of the potential and power of information warfare in the modern world, including information and communication technologies, sees it not only as a potential danger but also as a direct threat, by having the means of information warfare in the information sphere used against itself.

This book presents the postures and perception of the importance of several thinkers on military strategy and information operations such as Ionov,[79] Komov[80]and Gerasimov.[81] Although their work might be depicting the role of information-technical and information-intellectual activities in the Russian military thought sufficiently, it is appropriate for the sake of understanding the security and defense posture of Russia, to analyze official documents. The documents deal with national security, military strategy and defense in regard to information threat, information security and their relation to the strategic and doctrinal level. From the national security perspective, several documents are worth mentioning. These documents are "The Information Security Doctrine of the Russian Federation" from 2000,[82] "The Doctrine of information Security of the Russian Federation" adopted in 2016,[83] and the "Russian Federation National Security Strategy" from 2015.[84]

The following documents represent the military doctrine perspective: "Military doctrine of the Russian Federation" (2010),[85] "Conceptual views regarding the Activities of the Armed Forces of the Russian Federation in Information Space" (2010),[86] "The Value of Science is in the Foresight: New Challenges Demand Rethinking the Forms and Methods of Carrying out Combat Operations" (2013)[87] and "The Military Doctrine of the Russian Federation" adopted in 2015.[88]

The purpose of looking into the threat perception within these documents, and the aims they presented, is to tie together tools introduced and applied in the past and to show the reader that the importance of securitization of the information sphere, information itself and the processing of the data is of great significance to strategic documents today. Russia never changed its strategic mindset and thus, the West should not be surprised by the activities of Russia when it comes to striving for information superiority in the domestic and international scene. It is merely the re-enforcement of enacted policies and strategies with deep historical roots. The perception and role of information operations are ensconced in the Russian military and national security strategic thinking.

The documents analyzed below present the national security and defense posture to provide a strategic context for some notorious examples of information warfare utilization.

Direct excerpts of the documents constitute this chapter, so that the reader can have at hand the exact notation of the posture, instead of having to do further research. Author comments on the excerpts are directly in the text.

INFORMATION SECURITY DOCTRINE OF THE RUSSIAN FEDERATION

Valid as of 2000 to 2016, provided the readers with explicitly defined threats, a set of objectives and areas of thrust that, for the sake of the study, were merged into aims/objectives. The reader can extrapolate from the list of threats the variability and perspectives on the information sphere, information operations, influence, infrastructure, by the national security apparatus of the Russian Federation.

Threats:

"...activities of foreign political, economic, military, intelligence and information entities, directed against the interests of the Russian Federation in the information sphere."

This is a comprehensive conceptualization of threats. It deals directly with the malicious content and malicious code-based threats. Foreign information entities may consist of foreign media outlets – non-governmental entities operating in the information sphere – which today are almost all of them. Given that the interests of the Russian Federation defined in the policy directing documents are as broad as Russian traditional moral and spiritual values,[89] the threat perception and the domain where the threat is perceived, the information sphere, constitutes a doctrine-based frame for potential victimization.

"the striving of a number of countries toward dominance and the infringement of Russia's interests in the world information space and to oust it from external and domestic information markets"

The threat of dominance and infringement of interests by other countries points toward the direction of the stated importance of inherent information superiority in achieving strategic, operational and even tactical objectives in the political and military arena. This threat shows the defensive nature of the threat perception, although the means of information dominance in the global information space are being employed offensively by the proxies associated with the Russian government.[90] From the other perspective, nation states use digital protectionism in cases such as the U.S. stance on the Kaspersky antivirus program used at the U.S. government Federal agencies,[91] or U.S. approach to ZTE.[92] As noted in other documents, Russia perceives information and communication technology (ICT) dependence as a threat, and information and communications (hardware and software) independence as a strategic objective. Strategic autonomy for them means not having foreign manufactured and coded tools incorporated into Russian infrastructure.

"an increase in the technological edge of leading world powers and the buildup of their ability to hinder the creation of competitive Russian information technologies"

The menace of losing the ability to have a comparable technological edge with developed countries is in correlation to the past, when Soviet thinkers started to devise information warfare concepts aimed at decision-making in order to flatten the technological imbalance between the U.S. and the USSR. Also, the threat is presented in a defensive tone, as if the Russian inability to create genuine competitive information technologies is due to foreign influence. On the contrary, the issue might not be in the quality of the technologies or solutions, as Russian natural sciences, education institutions, mathematical and physics schools of thought are of the highest level and respected throughout the world.

"development by a number of states of information war concepts that provide for creating means for dangerous attack on the information spheres of other countries of the world, disturbing the normal functioning of their information and telecommunication systems, breaching the security of their information resources and gaining unsanctioned access to them;"

This constitutes a significant statement in which, apart from the evident and shared threat perception by other countries, such as attempts to access information and conduct computer network operations, Russia described the three layers of how it sees cyberspace. The division into information sphere, the infrastructure, and the information resources in this doctrine shows that there is an insufficient infrastructure-centric perception of cyberspace by U.S. and European proponents. Several other threat-associated statements below refer directly to the different perceptions of cyberspace between the Russian based model, and the U.S./European infrastructure-centric understanding of it.

"the immaturity of civil society institutions, and insufficient state control over the development of the Russian information market;"

The very fabric of civil society and the free flow of information, the right to free speech and the sharing of ideas that might be unfavorable to the ruling establishment, are addressed here. The understanding of this threat in the doctrine stands for content-oriented regulation and control in the information sphere, as is evaluating the civic society as an information resource, able to shape public perception.

"violation of the constitutional rights and freedoms of citizens that are realized in the information sphere;"

There is a broadened perception of what constitutes a threat here. However, judging by some of the arguments of the Russian establishment regarding the annexation of Crimea,[93] ideology such as this might serve as a pretext for launching information operations in the information sphere, if not in the physical domain.

"insufficient legal governance of relations in the area of the rights of different political forces to use the media for the advocacy of their ideas;"

This is threat perception based on content available in the information sphere. The language suggests that a legal framework should regulate the abilities of different political forces, either national or foreign, to campaign their ideas and preferences. Essentially, this would give the state the power to decide under what conditions it promotes a political agenda in the information sphere.

"the spread of disinformation about the policy of the Russian Federation, the activities of the federal bodies of state authority and events occurring in the country and abroad;"

Here, the threat perception is clear. Influence over the information resources, people, decision makers, and societal groups with information unfavorable to the Russian Federation and its interests, is based on the described concept of information dominance and superiority. Whoever achieves information superiority through the "reflection of reality to be believed," has considerable influence on the actual reality and events and behavior of other entities.

"activities by public associations, aimed at a forcible change of the foundations of the constitutional system and seeking to disrupt the integrity of the Russian Federation, foment social, racial, national and religious strife and spread these ideas in the media;"

Threat here derives from the content related aspect of information operations against the interests of Russian Federation in the information sphere, in this case, media.

"informational influence that foreign political, economic, military and information entities may have on the elaboration and implementation of the foreign policy strategy of the Russian Federation;"

Above is one of the key examples of the broad understanding of informational influence in Russia. As a tool of foreign entities on the elaboration and implementation of strategies and interests, informational influence is managed through various means. However, the political, economic, military or information categorization does not only define the entity, (international organizations, foreign governments, military, and media) but also defines the methods of influence operations in the information sphere.

"attempts at unsanctioned access to information or attack attempts against information resources and the information infrastructure of the federal executive bodies implementing Russian Federation foreign policy, of Russian representations and organizations abroad and the representations of the Russian Federation at international organizations;"

This threat is similar to what has already been presented, but serves as an example that the information sphere (as defined by Russia) is not only the cyberspace geographically associated with the Russian Federation, meaning information infrastructure and information sphere in Russia, but also encompasses the information spheres of foreign information infrastructure. It is of particular interest because of the proposed Code of Conduct.[94] The issue of sovereignty in cyberspace is an issue which could lead to the legitimization of censorship.[95] On the one hand, the Code of Conduct proposed by Russia, China, Tajikistan, Kazakhstan, Kyrgyzstan, and Uzbekistan intends for application of physical domain rights and sovereignty based on territorial integrity, but this particular threat assessment defines threat in the information sphere to Russian entities and interests way beyond the information sphere tied to Russian Federation geographically.

"the information and propaganda activities of political forces, public associations, media and individuals distorting the strategy and tactics in the foreign policy activity of the Russian Federation;"

In plain sight, this national security document describes the power of individuals conducting information and propaganda activities. It displays the gravity of how information activities are perceived by the Russian national security apparatus, and that the content of information is of high importance in preserving national security.

"insufficient provision of information to the public on the foreign policy activity of the Russian Federation;"

So far, the threats were aimed at actors or vectors perceived to be maliciously targeting Russia. Highlighting an insufficiently informed public indicates the need to provide content in the information sphere that opposes or refutes other sources the public might access. The objective is to provide the information sphere with the "correct" perception of reality and events.

"the creation of preferential conditions for foreign techno-scientific products in the Russian market and a simultaneous striving by developed countries to limit the development of Russia's techno-scientific potential (buying up shares of advanced enterprises with their subsequent refocusing, keeping export and import restrictions and so on);"

This threat again underscores the worries caused by perceived technological disparity, as well as the threat assessment of dependence on foreign technologies with influence from outside Russia, limiting Russia's potential in the field of ICT.

"a deformation of the system of mass information owing to media monopolization as well as to uncontrolled expansion of the foreign media sector in the national information space;"

An information sphere consisting of media and mass informatization is also of concern, as it enables open the dissemination of content to the public.

"foreign special services' use of media operating within the Russian Federation to inflict damage to the nation's security and defense capability and to spread disinformation;"

Here is another case where the Russian security apparatus, through threat perception, acknowledges the importance of disinformation impacting national security and even defense capability. With this position, there is the link to the utilization of these methods in favor of the Russian Federation, for example when it comes to operations with its own use of media for spreading disinformation.[96]

"the inability of contemporary Russian civil society to ensure the formation in the growing generation, and maintenance in society, of socially required moral values, patriotism and civic responsibility for the destiny of the country;"

The threat is not directly related to the information sphere but correlates with the perception of reality and information itself. A growing segment of civil society does not identify with the Russian establishment and thus constitutes a threat to the regime. The reaction is to provide the population with information – a perception of what are the correct, required, moral values and level of patriotism. In a nutshell, it is the spread of information to form the perception of reality and the creation of a base for control of the narratives – what is good for the country. The reflexive control principle of subject-object relationship comes into play.

"informational and technical influences (including electronic attacks, penetration into computer networks) by likely adversaries;"

This is the sum of all fears: a combination of activities in the realm of content and code based operations, including attacks on the information sphere psychologically, information resources themselves and the utilization of technical, code based information weapons.

"subversive and sabotage activities by special services of foreign states, carried out by methods of informational and psychological influence;"

Psychological influence adds to the informational and technical influences, which makes the reflexive control even more effective.

"possible information and propaganda activities undermining the prestige of the Russian Armed Forces and their combat readiness;"

The last threat presented in this analysis is no less striking. Descriptions of information and propaganda activities as a threat to prestige – the perception of status or role in society for example – and influencing combat readiness shows that the informational base of the threat has direct implications on the ability to stand firm/carry out the mission in the case of a military operation. It encapsulates the importance of information and propaganda and their roles in the national information security posture of the Russian Federation.

The sources of threats presented in the document mirror Russian information warfare concepts regarding information-psychological and information-technological weapons. The aim is not only to secure the information and protect the infrastructure (as in the U.S./European infrastructure-centric concept of cybersecurity) but also factors in altering the content and dealing with the spread of material unfavorable to the interests of the Russian Federation. The threat perceptions presented in the Information Security Doctrine document are good examples of how Russia sees content as a source of power or a tool that can be used to impact national security and defensive capabilities. Also, all the presented threats – which are projected by different actors – are not limited to the time of armed conflict. Throughout the entire document, there is no distinction made between peacetime and "wartime." This also shows that the lines are blurred and that Russia does not take them into account when dealing with those threats. The threats are constant and not exclusive to a specific phase of a conflict.

The threats from a document adopted in 2000 were relevant then. However, as presented later in the chapter, they are still appropriate for the Russian Federation, and present in national security and defense doctrines today.

The year 2000 precedes the Russian operational campaigns in Estonia, Georgia, and Ukraine. In the cases of Georgia, and especially Ukraine, many analysts seemed to be flabbergasted by the "new" approach to how combat operations were conducted, citing the employment of new hybrid warfare. Also, how a military campaign is complemented with information warfare using all means available (civilian, military, and diplomatic), was the subject of debate and surprise. By reading official documents and familiarizing ourselves with threat perceptions, and later in the paper with the objectives and aims, the analytical community could have been better equipped to understand the measures and the toolbox employed by the Russian forces and stakeholders in the conflicts. Russia was well aware of the technological edge of other countries. The author repeatedly mentioned the disparity in technology. So the Russians, as good strategists, set off to exploit the weaknesses available and deploy the assets available to counter the edge.

The doctrine also defines aims, areas of thrust, leading to objectives. They are presented below with commentary.

AIMS/OBJECTIVES:

"developing the theoretical and practical foundations of national information security assurance with regard for the current geopolitical situation, Russia's political and socioeconomic development conditions and the reality of the use of the "information weapon;"

Here is an overall objective which encompasses the organizational and institutional framework for assuring information security, the legal framework and regulation tools aiming at ensuring information security, to concepts and assets to provide information security, such as how to face threats from the use of information weapons. It also includes the development of those same things. It is imperative to bear in mind the Russian perspective on information security[97] is broader that the U.S./European concept of it.

"securing the technological independence of the Russian Federation in the major areas of informatization, telecommunications and communication determining its security, and primarily in the field of developing specialized computer hardware for weapon and military equipment specimens;"

Above is a direct reaction aimed at the threat of being dependent on foreign technologies in the national security and military complex. The Russian establishment is well aware of the risks of implementing foreign technology. The previous negative experiences were too significant to be omitted.[98]

*"crafting and adopting normative legal acts of the Russian Federation to establish legal and natural persons' responsibility for unsanctioned access to, and the illegal copying, distortion or unlawful use of information, **the deliberate circulation of untrue information**, the*

illegal disclosure of confidential information and the use of business or trade secret information for criminal or ulterior purposes;"

The question regarding this objective is, who is to define misleading information, and how? Apart from the "untrue information" factor, the deliberate circulation aims at the information sphere.

"making more precise the status of foreign news agencies, media and journalists as well as of investors when attracting foreign investment for the development of Russia's information infrastructure;"

The objective from the year 2000 was a subtle kick-off towards the bill adopted in 2017 in the Duma.[99] The law forces foreign journalists and NGOs to undergo scrutiny, leading to judicial punishment for the non-compliant. The bill is seen as a tool for the establishment to gain increased control over the civil society and its presence in the information sphere.[100,101]

"determination of the status of organizations providing the services of global information technology networks within the territory of the Russian Federation, and the legal regulation of the activities of these organizations;"

The expert has the same objective as listed above, but is aimed at the technical and social network side of information space. It is intended not only for potential foreign agents, but also for the domestic arena. For example, in 2014 the Russian social network Vkontakte was efficiently put under control of the Russian government apparatus[102] and its founder Pavel Durov faced mounting pressure to provide information on its users, something with which he refused to comply. This is an example of the establishment leaning towards greater control of the information sphere regarding not just services, but also content and access to information of its users.

"creation of systems and means for preventing unsanctioned access to information being processed and special attacks causing the distortion, damage or destruction of data as well as alteration of the normal operating modes of informatization and communication systems and means;"

The above objective is based on the need to provide the national security apparatus with technical and organizational means for intrusion detection and intrusion prevention systems, probes and the ability to monitor the networks and flow of information. It also aims at hardening the networks to limit the potential penetration and alteration of their functionality, making those networks more difficult to compromise.

"formation of the system of monitoring the indicators and characteristics of the information security of the Russian Federation in the most

important spheres of life and activity of society and the state;"

This objective looks subtle and reiterates the previous one with a slight difference; there are no descriptions of indicators and characteristics of information security, which also encompasses not only infrastructure security, but also users and content. The materialization of this objective could be the SORM-3,[103] based on a ministerial directive from 2014, ordering the telecommunications operators to install equipment compliant with mandated technical requirements, which allow targeted surveillance and wiretapping. The SORM-3 enables for deep packet inspection.

"establishing a system for countering the monopolization of components of the information infrastructure by domestic and foreign entities – including the market for information services and the mass media;"

The system for countering monopolization serves in this objective as a cover for not being under control or direct influence of the establishment.

"stepping up counterpropaganda activities aimed at preventing negative consequences of the spread of disinformation about Russian domestic policy;"

The objective was fulfilled partly by the foundation of Sputnik[104] and Russia Today,[105] later RT, a Russian television network funded by the Russian government.[106] The broadcast scope is international and provides viewers with official positions favorable to the Russian government on foreign and domestic issues. RT serves as one of the leading propaganda channels. The objective is presented explicitly as defensive, to provide alternative perspectives and perceptions of reality differing from those unfavorable to the Russian establishment. However, RT is involved in spreading disinformation itself.[107]

"the creation for Russian overseas representations and organizations of conditions for work on the neutralization of the disinformation being spread there about the foreign policy of the Russian Federation;"

It is a clear objective aimed at influencing the perception of reality through creating a false sense of authority via think tanks, fora for attracting willing experts and politicians and institutions to legitimize disinformation.[108]

"perfecting the information support of the work on counteracting abuses of the rights and freedoms of Russian citizens and legal entities abroad;"

The vague objective aims at the creation and elaboration of information support to promote the interests of the Russian Federation masked as counteracting the abuses of rights and freedoms of its citizens. Information support stands for not only creating content itself intended for dissemination and related guidance, but also nurturing of channels for propagation (individuals, networks, organizations).

"structural improvement of the functional agencies of the information security system in the defense sphere and coordination of their mutual activities;"

Apart from the tools and methods to be implemented, researched and adopted, the fast pace of development in the information sphere and the increasing speed of events internationally calls for better coordination and adjustment of organizational bodies. The Russian national security apparatus is aware of this, and in 2000 devised the objective of improving existing structures.

"international efforts in prohibiting the development, spread and use of the 'information weapon';"

The aim was partly fulfilled by the proposal of Code of Conduct to the United Nations, where Russia, Tajikistan, China and Uzbekistan proposed among other things "Not to use information and communications technologies, including networks, to carry out hostile activities or acts of aggression, pose threats to international peace and security or **proliferate information weapons** or related technologies."[109]

However as there is no international consensus regarding exactly what constitutes an information weapon, and as the Russian concept of it is sufficiently broad to encompass even social networks in some instances, the term was omitted from the 2015 second proposal of the Code of Conduct. This omission was perhaps to gain a higher degree of support from the international community.

"securing the international exchange of information, including information flows via national telecommunication and communication channels;"

The aim here is blunt: to have the information infrastructure under the control of the government and to not allow private foreign entities to own a crucial part of the information sphere.

"determining the legal status of all parties to relations in the information sphere, including information and telecommunication system users, and establishing their responsibility for the observance of the Russian Federation laws in this sphere;"

The objective above in its broadness provides for recodification of terms and conditions – the rights, and obligations of entities involved in using, producing, maintaining and securing the information sphere. The determination process may lead to unforeseen circumstances imposed by the establishment to gain more control in the sphere.

"the creation of a system for the gathering and analysis of data on sources of threats to the information security of the Russian Federation, and on the implications of their accomplishment;"

Paraphrased as creating organizational, legal, technical and nontechnical means to protect the interests of Russia's information security. Given the broad meaning of information security, this statement is related to code, content, influence, and surveillance.

"improvement of the ways and means of providing strategic and operational camouflage and conducting intelligence and electronic countermeasures, along with the betterment of methods and tools for actively countering propaganda, information and psychological operations by a likely adversary;

The aim is transparent and requires no further commentary, except one reminder: the doctrine is a civilian document, although the methods are usually associated with military activities and armed combat operations.

"developing the infrastructure of Russia's unified information space, counteracting information war threats in a comprehensive way;"

Since 2000, the Russian establishment has been striving to toughen its control over the information sphere, with the culmination of plans to create Russia's own so-called internet; in fact, a unified information space under the control of Russia and other like-minded countries.[110]

"creating secure information technologies for systems used in the course of the realization of the vitally important functions of society and the state, suppressing computer crime, devising a special purpose information technology system in the interests of the federal bodies of state authority;"

This last objective marks again the intent to build particular purpose information systems, as in the case of SORM-3 for surveillance, but also others of an unspecified nature.

The reader should not be surprised by the information operations conducted domestically and internationally, either directly by the Russian establishment or indirectly through proxies. The doctrine laid out the threats and objectives in securing the interest of the Russian Federation. By reading through documents like this in 2000, the readers might have been acquainted with what Russia was fearful of, and thus realized to what extent they might exert themselves to protect their interests. Their threat assessment and proposed counter activities gave away the understanding, methods, and tools they were about to employ in not so distant future.

In 2000-2017 the Russian Federation was not doing anything unexpected in the realm of information security. Some of the activities might have been anticipated. The combination of past tools and methods of influence, modern technology and implementation of publicly presented strategies, with the declaration of perceived threats and actions how to tackle them, was like delivering the strategic means and intent on a silver plate.

As stated by Sergei Medvedev, one of the most striking divergences from the U.S. and European perception of the role of media is the Russian "illiberal attitude." He continues, "Regardless whether a media entity is private or state-owned, the doctrine states that it is acceptable and essential that the government ensures pro-Russian messaging. The Russian government clarified that this position intended only to provide state oversight and not censorship."[111]

In 2016 the "Doctrine of Information Security of the Russian Federation" replaced the "Information Security Doctrine" of 2000. Again a set of threats and objectives are presented with a commentary.

DOCTRINE OF INFORMATION SECURITY OF THE RUSSIAN FEDERATION (2016)

Threats:

"While a wider use of information technologies contributes to economic development and

better functioning of social and State institutions,
it also gives rise to new information threats."

The first threat vaguely formulates that broad
use of technology enables new information
threats. The question is whether this is just stating
the obvious or the authors had in mind information threats that were present with lesser
penetration of information technologies in
everyday life and are warning against new, as yet
unidentified ones.

"The possibilities of transboundary infor-
mation circulation are increasingly used for
geopolitical goals, goals of a military-political
nature contravening international law or for
terrorist, extremist, criminal and other unlawful
ends detrimental for international security and
strategic stability."

The critical takeaway terms of this threat perception are "transboundary nature of information
circulation," "military-political nature" and
"strategic stability." Each of them provides the
readers with a better understanding of the
Russian national security mindset. The transboundary nature of information circulation is
against the interest of the Russian Federation just
because Russia strives to have information
resources and channels of distribution under
control and regulated by the security apparatus.
The danger of having an uncontrolled circulation
of information from foreign entities crossing into
the Russian information sphere with no re-

strictions or ability to influence constitutes a threat to the perception of reality and interests pursued by the Russian Federation. Military-political nature provides the readers with a better picture of how much the political and military tools are intertwined, and how small the differences are between the toolboxes used in pursuit of strategic objectives. The last term – strategic stability – refers to the Cold War era in the sense of a strategic standoff due to nuclear capabilities of the U.S. and the Soviet Union. The notion of it is essential because it first refers to the superpower status which Russia attempts to maintain, and secondly because it ensures status quo in the international order.

"Moreover, the practice of adopting information technologies without due consideration of their impact on information security significantly increases the probability of information threats."

The national security apparatus is well aware of threats coming from a precipitous deployment of technologies without consideration of the effects in security-related matters. Sometimes convenience and pursuit of efficiency trump security considerations.

"One of the key negative factors affecting the state of information security is the fact that a number of foreign countries are building up their information technology capacities to influence

the information infrastructure in pursuing military purposes."

The threat perception summarizes and reiterates the threats seen in the doctrine from 2000.

"Intelligence services of certain States are increasingly using information and psychological tools with a view to destabilizing the internal political and social situation in various regions across the world, undermining sovereignty and violating the territorial integrity of other States. Religious, ethnic, human rights organizations and other organizations, as well as separate groups of people, are involved in these activities and information technologies are extensively used towards this end."

This statement refers to threat towards the Russian Federation by foreign influence and to an international order preferred by the Putin administration. However, the same danger that destabilizing the internal political and social situation presents, the undermining of territorial integrity and sovereignty, are mirror examples of what Russia was and is doing in several cases itself: Georgia, Ukraine, and Moldova. Russia sees the employment of these very same methods and tools described as a threat towards itself, especially knowing how effective those tactics are.

"There is a trend among foreign media to publish an increasing number of materials

containing biased assessments of State policy of the Russian Federation."

The above notion signifies the defensive and almost hurtful assertion that the external actors are defiling the Russian Federation policies.

"Russian mass media often face blatant discrimination abroad, and Russian journalists are prevented from performing their professional duties."

The threat, according to the author, is not a direct threat to the information security of the Russian Federation, links up with two examples. First is the U.S. government's pressure towards the RT to register as a foreign agent[112] after the U.S. intelligence agencies reported that RT is a Russian, state-run, propaganda machine. Second is the mounting pressure from intelligence communities that affiliated Russian journalists are abusing their press status and are conducting intelligence activities on behalf of the Russian Federation.[113]

"There is a growing information pressure on the population of Russia, primarily on the Russian youth, with the aim to erode Russian traditional spiritual and moral values."

Here is a previously described perceived threat of malicious content aimed at the population of Russia, hiding the threat of officially unsanctioned material behind the traditional and moral values fabric.

"Information security in the sphere of State and social security is characterized by a continued increase in the complexity, scope, and coordination of computer attacks on objects of critical information infrastructure, enhanced intelligence activities of foreign States against the Russian Federation, as well as growing risk that information technologies will be used to infringe on the sovereignty, territorial integrity, or political and social stability of the Russian Federation."

First, it may be noted that the authors of the doctrine perhaps wanted to describe the information insecurity. Also, yet again this is an example how serious information security is in the realm of content and code, when it has the potential to influence the sovereignty, territorial integrity and social and political stability of Russia. Again, the authors are well aware of the power and potential impact of a well-conducted information operation.

"Information security in the economic sphere is characterized by a lack of competitive information technologies and the inadequate use of information technologies in the production of goods and services. The level of dependence of the domestic industry on foreign information technologies, such as electronic components, software, computers and telecommunications

equipment remains high, which makes the socioeconomic development of the Russian Federation dependent on the geopolitical interests of foreign countries."

This is a perceived threat to national security that had been present in the strategic thinking of the Soviet Union, and occupies the national security arena even today – the Russian dependence on foreign technologies and the technological disparity. On the one hand, the concern is understandable. However, it almost sounds like an excuse that the geopolitical interests of foreign countries are behind the lack in the production of high-end technology in Russia. As mentioned previously, in general the human potential in mathematics, physics, and natural sciences, is sufficiently present in Russia. The natural resources of the Russian Federation also offer a sufficient base for high tech industry, and the spread of technologies and knowledge in recent decades provided for a solid knowledge base to jump start research and development. Attribution of the failure of the Russian Federation to develop its technologies can hardly be associated with the geopolitical interests of foreign countries. There might be nevertheless a historical-political explanation of the lack in computerization of the Soviet and early Russian Federation society. Timothy Thomas offers the following thought:

> In the former Soviet Union, comprehensive integration of computer technology was delayed by two factors: the Soviets' iron grip over most information and tech-

nology systems (Xerox machines, personal computers, patents, etc.); and a reluctance to study information systems seriously. In fact, the West had an early unchallenged entry into the computer age since cybernetics was officially proscribed by General Secretary Nikita Khrushchev only during the late 1950s.[114]

"Information security in the sphere of strategic stability and equitable strategic partnership is characterized by the desire of individual States to use their technological superiority to dominate the information space."

This threat closes the list and refers directly to strategic stability, which provides for the balance between superpowers, or today, developed states. It points mainly at the U.S., seen as the primary opponent of Russia in spreading its narratives and information campaigns in the information sphere. The information sphere is therefore regarded as a strategic asset reaching the significance of nuclear arsenals in the past – solely for the ability to alter strategic stability on the global scale. This is the perception of the Russian strategic thinking.

The Doctrine also defines strategic objectives, trusts, and critical areas. They are presented below with a commentary. In the case of this Doctrine, they are divided by areas of responsibility, such as information security in the field of national defense, state and public safety and field of strategic stability and equal strategic partnership among others.

OBJECTIVES/AIMS:

National defense domain

"The military policy of the Russian Federation identifies the following key areas of ensuring information security in the field of national defense: ensuring strategic deterrence and preventing military conflicts that may be brought about by the use of information technologies;"

In the past, strategic deterrence referred to nuclear arsenals from the U.S./European perspective. However, "strategic deterrence in the perception of Russia is a doctrinal approach built on a demonstrated spectrum of capabilities and a resolve to use military force."[115] It is not limited to nuclear weapons. "It is conceived much more broadly than the traditional Western concept of deterrence. It is not entirely defensive: it contains offensive and defensive, nuclear, non-nuclear and non-military deterrent tools. These are to be used in times of peace and war – making the concept resemble, to Western eyes, a combined strategy of containment, deterrence and coercion – using all means available to deter or dominate conflict."[116]

It also encompasses information weapons, information struggle and the critical non-military tools, such as influence operations, to undermine decision-making algorithms to achieve dominance in the conflict. Information superiority is a necessary part of the process for attaining and maintaining strategic deterrence. To meet these ends, information warfare and information technology are of high importance.

"upgrading the information security system of the Armed Forces of the Russian Federation, other troops, military formations and bodies, including forces and means of information confrontation;"

Here is a clear statement of the necessity to enhance the information weapons and concepts to improve the ability to strike in the information sphere and conduct information operations.

"forecasting, identifying and assessing information threats, including threats to the Armed Forces of the Russian Federation in (the) information sphere;"

This objective relates to the ability to conduct analysis, and to conceptualize threats emanating from the information sphere. It is important to be able to adopt new concepts of information warfare, to either keep the information superiority or achieve it when necessary.

"promoting the interests of the Russian Federation's allies in (the) information sphere;"

It is interpreted as international support in the information sphere, and the promotion of the interests of Russia and its allies. It shows the aspirations to influence for the sake of their alliances other entities, to which Russia has a reserved position as it identifies the same aim as a threat against itself.

"countervailing information and psychological actions, including those aimed at undermining the historical foundations and patriotic traditions related to defending the homeland."

The objective above refers to countering activities seen as hostile and of an informational and psychological nature. It is covered by mentioning historical foundations, along with a sprinkle of patriotism. However, the aim is clearly to counter any activities that differ from the official perception of the Russian establishment. Interestingly it is mentioned in both sections, the national defense section and the state and public security domain. Again the lines between responding and the deployment of military means and civilian security measures are blurred.

State and public security domain

"countering the use of information technologies to promote extremist ideology, spread xenophobia and ideas of national exceptionalism for the purposes of undermining the sovereignty, political and social stability, forcible changing the constitutional order and violating the territorial integrity of the Russian Federation;"

This objective reiterates already stated threats. The repetition of it across documents, persistence in time, and presenting it in different sections of security and defense domains, shows how vital and feared the potential use of information technologies is. It occupies the top levels of consideration when it comes to preserving the sovereignty and territorial integrity, and accordingly, it shows the level of attention it bears in Russian strategic and military thought.

"suppressing the activity detrimental to the national security of the Russian Federation, carried out by special services and organizations of foreign States as well as by individuals using technical means and information technologies;"

The interesting fact about this objective is the differentiation between technical means and information technologies, as if information technologies had a broader meaning in the Russian line of thought.

"enhancing the safe operation of information infrastructure objects, including with a view to ensuring stable interaction between government bodies, preventing foreign control over these objects, and ensuring the integrity, smooth operation and safety of the unified telecommunications network of the Russian Federation, as well as ensuring the security of information transferred through this network and processed within information systems in the territory of the Russian Federation;"

The objective has a link to the "Russian internet" regarding the unified telecommunications network of the Russian Federation. Also, it stresses not only unification of telco network, but also the geographical aspect, which is an underlying argument in the discussion about states' sovereignty in the information sphere as a state, in the proposed Code of Conduct.

"neutralizing the information impact intended to erode Russia's traditional moral and spiritual values."

Once more, the object refers to the vaguely defined moral and spiritual values of Russia, to be enforced quickly. How? Any content that is deemed by the establishment as unfavorable to it is to be labeled extreme or unpatriotic, or harmful to Russian interests. "Neutralization" is a broad concept; is it going to be done through counter-narratives, or through legal frameworks leading to judicial punishment? Business-related sanctions? Anyway, the erosion of values, the nation's narrative and cohesion is of great importance, because if exploited, it has a significant impact on the society's ability to govern itself.

"eliminating the dependence of domestic industries on foreign information technologies and information security means by creating, developing and widely implementing Russian solutions, as well as producing goods and providing services based on such solutions;

Reiterated across the documents, the threat of technological dependence and realizing the technological disparity leads to the government's push for technological self-sustainability in the information sphere. The information market is also preferred to be of Russian origin, or controlled by Russian companies, if not by the establishment itself.

"protecting citizens from information threats, including by promoting the culture of personal information security."

It is the only objective for individual security, aimed towards personal information security across the documents; otherwise it is about infrastructure, content, systems, values.

The role of individuals is underestimated as the weakest link, and yet, the document mentions the potential for education in the sphere of personal information security only once, and vaguely at that.

"protecting the sovereignty of the Russian Federation in information space through nationally-owned and independent policy to pursue its national interests in information sphere;"

The above objective again confirms the direct link between information security and sovereignty in the information sphere.

"taking part in establishing an international information security system capable of effectively countering the use of information technologies

for military and political purposes that are contrary to international law, or for terrorist, extremist, criminal or other illegal purposes;"

The statement relates to the proposed Code of Conduct and the work between members of the Shanghai Cooperation Organization (SCO) aiming at codification of use of information-based concepts in the realm of political and military objectives.

"developing a national system of the Russian Internet segment management."

And last, but not least, there is a subtle line referencing the Russian Internet.[117]

"The Russian Federation National Security Strategy" from 2015 seems to be more politically shaped in defining the threats not as systemic, or organizational, but directly naming Russian rivals. It starts with asserting that the application of independent policies by Russia leads to suppression by the United states and its allies. It directly states the pressures adopted to do so:

*"...political, economic, military, and **informational pressure** on it."*[118]

*"An entire spectrum of political, financial-economic, and **informational instruments have been set in motion in the struggle for influence in the international arena."***

The strategy, in light of the 2013 and 2014 and ongoing events in Ukraine, is asserting that as an example of this informational pressure we are witnessing:

"the deliberate shaping in the Ukrainian population of an image of Russia as an enemy..."

Considering the activities of Russian armed forces and annexation of Crimea and ongoing hostilities with many casualties, this seems unnecessary.

> The intensifying confrontation in the global information arena caused by some countries' aspiration to utilize informational and communication technologies to achieve their geopolitical objectives, including by manipulating public awareness and falsifying history, is exerting an increasing influence on the nature of the international situation.[119]

The strategy postulates that Russia is developing and implementing:

*"**Interrelated** political, military, military-technical, diplomatic, economic, **information-al**, and other **measures are being developed and implemented in order to ensure strategic deterrence** and the prevention of armed conflicts."*

In the sum of all documents, whether information security doctrines, security strategies or military doctrines, this is a blunt summarization of the efforts and the tendency to use and exploit information measures to achieve and retain strategic deterrence. An important aspect is the notion of armed conflict prevention. The objective here is to employ assets as mentioned earlier, in order to reach the strategic goal without reaching the threshold of armed conflict by international law. Accountability to the international community is limited, and it allows for tactical and strategic efforts to steer clear of a military conflict as defined by international humanitarian law.

The rationale behind not reaching the threshold by the above-mentioned means is clear: non-military application is efficient, in that each of the actions conducted through political, economic or informational measures, in and of themselves, do not meet casus belli criteria; however, their totality adds up to the achievement of a strategic objective. The advantage, among others, is the low cost of entry, in comparison with a large-scale conventional military campaign. More important-ly, it is hard for the targeted entity to react to a wide range of hostile activities – the combined tools of the whole statecraft spectrum.

"measures are being taken to increase the protection of citizens and society from the influence of destructive information from extremist and terrorist organizations, foreign special services, and propaganda structures;"

It is an essential role of the state to protect its citizens from extremist and terrorist organizations

– not only from kinetic attacks, but also from spreading content leading to further radicalization or recruitment. However, what are the propaganda structures mentioned in the paragraph of the National Security Strategy? It is discussed separately from foreign special services, namely its intelligence community, separately from extremist and terrorist organizations? Among the aims of the strategy, one is outstanding when it comes to the informational and psychological means of influence.

*"ensuring of the **Russian Federation's cultural sovereignty** by means of taking measures to protect Russian society **against** external expansion of ideologies and values and **destructive information and psychological impacts, the implementation of control in the information sphere**, and prevention of the spread of extremist products, propaganda of violence, and racial, religious, and interethnic intolerance;"*

The objective refers to the control in the information sphere, an external threat that Russia would like to itself control via the same methods used to describe the threat.

This notion of control in the information sphere, described by military theorists and strategists as a very important prerequisite, correlates with those concepts of information superiority, information operations, influence campaigns and reflexive control consistently outlined in their governing strategy documents.

The countermeasures proposed for external threats and enemies are employed and utilized by the establishment to reinforce itself.

The Strategy concludes with a declaration:

"In implementing this Strategy, particular attention shall be paid to ensuring information security in light of strategic national priorities."

MILITARY DOCUMENTS

The role of the information sphere, information warfare, modern conflict and features of modern technological conflicts are also covered in the military doctrines and subsequent documents of the Russian Federation.

Four documents will be presented in this chapter, namely the:

- Military Doctrine of the Russian Federation (2010)
- Conceptual views regarding the Activities of the Armed Forces of the Russian Federation in Information Space (2010)
- Value of Science is in the Foresight: New Challenges Demand Rethinking the Forms and Methods of Carrying out Combat Operations (2013)
- Military Doctrine of the Russian Federation adopted in 2015.

MILITARY DOCTRINE OF THE RUSSIAN FEDERATION (2010)

Russian military doctrine[120] sets off with the proclamation to place information instruments at the same level as legal, political, military and others to *protect its interests*.[121] Among those instruments, as we have learned, are hostile code and hostile content, information superiority, control of the information sphere, reflexive control and other measures adopted in the analog past and newly reintroduced in the era of modern technology.

The military doctrine describes features of contemporary and modern conflicts. The aim is to provide perspectives on Russian military thought and the importance of elements for conducting operations to protect the interests of the Russian Federation. Of particular note to this study is this phrase in the doctrine: *"intensification of the role of information warfare."*

This is particularly true in relation to the growing importance of the speed and pace of command and control. As stated in the doctrine: "an increase in the promptness of command and control" is tied to the fact that today all militaries are applying modern technologies. This does not stem only on the technological development level, but also leaves no choice for them other than to abandon the "strict vertical system of command and control to a global networked automated command and control system for troops (forces) and weaponry;"

The transition they seek enables better agility, understanding of the battlefield, and visibility, by ingesting information and command. However all these activities which are essential during combat are also fragile – not only because they are part of the decision-making algorithm, but because the networks and technologies used are susceptible to either hostile code or content.

The dependence on technology and a decision-making system which rely on vast quantities of relayed, stored and processed information, are also exploitable features in contemporary conflicts. The vulnerabilities of modern decision-making systems and information technologies, sensors, databases and communication channels are ideally suited to the development of concepts to achieve tactical, operational and strategic objectives in the political or military arena.

The doctrine elaborates on the features of modern military conflicts, offering a glimpse into strategic thinking regarding the position of information warfare in the contemporary military thought. The elaboration on the nature of military conflict sets the stage for subsequent tasks which were intended to be transitional. With the era of mass mobilized Soviet armies over, the missions and skill-sets for current and future military conflict required a different approach. With that context, the document claims a defining feature: "...prior implementation of measures of information warfare in order to achieve political objectives without the utilization of military force and, subsequently, in the interest of shaping a favorable response from the world community to the utilization of military force."

Accordingly, nonmilitary tools and methods for influencing the general public, and also the international community – and especially opposing decision makers – will be deployed and used, either to achieve established goals or prepare the scene before actual deployment of forces. There can hardly be a more explicit statement in the official documents that opponents of the Russian Federation will experience information warfare while Russia seeks its objectives. The pursuit of the strategic interests of the Russian Federation has a predecessor; information warfare.

To fully understand the extent of the role of information warfare with all its attendant features (such as technical and intellectual, use of hostile code and hostile content, targeting of communication channels, the information sphere, information resources and the decision-making system), the following features of modern conflict from the doctrine are presented: a dependence on advanced technology, undisrupted transfer of information, swift and precise decision-making.

Feature 1:

"Military conflicts will be distinguished by speed, selectivity, and a high level of target destruction, rapidity in maneuvering troops (forces) and firepower, and the utilization of various mobile groupings of troops (forces)."[122]

The tempo of a military conflict translates from the ability to rapidly move your forces and to use the element of surprise to your advantage. A high level of target destruction is achieved not only by modern weaponry and ammunition, but also with technology reliant on target acquisition, calculation, and navigation. All those aspects are heavily based on information infrastructure, communication channels, encryption, and the availability of signal, resistance to jamming and spoofing of satellites, among others. Also, the decision-making process for targeting and command, in general, is entirely dependent on the ability to obtain proper information, trust in the fidelity in the information provided, and the ability to process and properly store that information for future combat operations. The psychological aspect of decision maker bias also comes into play.

Feature 2:

"Possession of the strategic initiative, the preservation of sustainable state and military command and control, and the securing of supremacy on land, at sea, and in the air and outer space will become decisive factors in achieving objectives."[123]

A strategic initiative is not only about surprise, operational art and the ability to escalate and de-escalate freely without being tied to a reactive position. In the present world, it is also about setting the narrative for the general public, international community and decision makers. The ability to do so is vital for achieving objectives without being caught in a compromising position, should the acts ultimately be deemed unlawful. Attaining strategic initiative also sows distrust, debilitates the decision-making processes of the opponent, and seeks to destroy the societal cohesion within the target entity.

Feature 3:

"Military actions will be typified by the increasing significance of precision, electromag-

netic, laser, and infrasound weaponry, comput-
er-controlled systems, drones and autonomous
maritime craft, and guided robotized models of
arms and military equipment."[124]

Above-mentioned weaponry, including new physical and kinetic features of such technologies and weapon platforms, depend heavily on computer-based technology, information and communication channels. Computer supervised systems, entirely controlled by algorithms, are the dawn of a new era in warfare, but also evidence of a new set of exploitable inherent vulnerabilities, such as the takeover of automated systems for perimeter control, lethal systems, surveillance assets or even whole swarms of automated or semi-automated military platforms.

These three features have a few things in common – the reliance on technology, decision-making algorithms and the link to all three levels of command and control – tactical, operational and strategic.

To deter and prevent military conflict, the Doctrine presents more than a few tasks:

- *"to assess and predict the development of the military-political situation at the global and regional level and also the state of interstate relations in the military-political sphere **utilizing modern technical systems and information technologies;**"*[125]

- *"to improve the system of information support for the Armed Forces and other troops;"*

- *"to develop forces and resources for information warfare;"*

- *"to create **new** models of **high-precision weapons and develop information support** for them;"*

- *"to create basic **information management systems and integrate** them with the **systems for command and control** of weapons and the **automation systems of command and control organs at the strategic, operational-strategic, operational, operational-tactical, and tactical levels.**"*

The "Military Doctrine of the Russian Federation" was published in 2010; it was followed by the "Conceptual Views Regarding the Activities of the Armed Forces of the Russian Federation in Information Space" document.

The reasons behind the publication of this document are unclear. Perhaps it is to further elaborate on the topic of activities in the information sphere or to stress even more the importance of informational concepts in modern warfare. The document looks at the definitions from an "information-centric perspective, defining information war as actions that may damage information systems and resources; undermine political, economic, and social systems; brainwash the population; or coerce the victim government."[126]

Also, the Conceptual document states that information space is another operational domain. This statement predates by six years NATO's declaration of cyberspace as a functional domain at the Warsaw Summit in 2016.[127] "Along with the land, sea, air and outer space, the information space has been extensively used for a wide range of military tasks in the armies of the most developed countries."

The conceptual document postulates that information weapons have cross-border effects due to the ubiquitous nature of information technologies, and so the role of information warfare has gained greater importance.[128]

"The Doctrine lays down the priorities to counter this threat. One of them is to improve techniques and methods of strategic and operational deception, intelligence and electronic warfare, methods and means of active information and psychological operations counter measures. Besides, since recently computers have been widely used in command and control and weapons control systems, that's why the list of threats is now supplemented with the task to protect the information infrastructure of the Armed Forces of the Russian Federation against various types of computer attacks."

The conceptual document reveals several definitions that so far were either presented by military thinkers or derived from context and its text offers official elaborations of these terms. First and foremost is the definition of military conflict in information space:

"Military conflict in the information space is a form of interstate or intrastate conflicts with the use of information weapons."

No less important is the definition of information war:

Information War is the confrontation between two or more states in the information space with the purpose of inflicting damage to information systems, processes and resources, critical and other structures, undermining the political, economic and social systems, a massive psychological manipulation of the population to destabilize the state and society, as well as coercion of the state to take decisions for the benefit of the opposing force.

By inflicting damage to essential information resources, coercing the opposing state by using information campaigns, using hostile code and content to alter decisions in favor of the attacker, the controlling subject aims to keep hostilities below the threshold of armed conflict, and below the threshold of regulations defined in international humanitarian law. The benefit here is that the target would be precluded from being able to employ conventional countermeasures, and is left struggling to assess the real degree of hostilities and how to respond.

Command and control, the military equivalent of decision-making in the civilian sphere of governance, plays a crucial role in the ability to conduct military operations, carry out strikes or merely coordinate and function as an entity with a hierarchical structure. Therefore, every structure authorized to carry out defensive and offensive operations strives to harden and enhance the resiliency of its infrastructure, processes and the commanders. Likewise, disrupting an enemy's command and control is a way to efficiently debilitate the opposition during the conduct of their operations, sometimes without conventional hostilities. Therefore, the Doctrine also states the following:

"In the conditions of the information warfare, adopting strategies to protect information resources will allow avoiding the disorientation of military command structures, C2 disruption, irreparable destruction of logistic and transport infrastructure elements, psychological disloca-tion of personnel and non-combatants in a war zone."[129]

As mentioned above, the military entity strives not only to protect itself and the command structures in general from attacks, but also to adopt such measures to conduct these type of attacks against potential adversaries.

"At the same time, the need for adoption of such measures on a priority basis in the current context is due to but not limited to the fact that hundreds of millions of people (whole countries and continents) are involved in a single global information space formed by the Internet, electronic mass media and mobile communication systems."[130]

The extent of the activities in information space is well described in the Doctrine as well. Yet again it shows a broad perspective on information security shared among Russian military thinkers and strategists:

"In general, operations in the information space are comprised of the staff and field intelligence efforts, operational deception, electronic warfare, communications, code and automated C2, information work of HQs, as well as protection of friendly information systems against electronic, cyber and other threats."[131]

The above perspective effectively presents the number of opportunities at the disposal of the adversary to disrupt command and control. However, it is not reserved exclusively for the highest echelons of command and control. As the Doctrine points out, the organization of the operations in information space is pertinent to all levels of leadership and all times, peace or war.

"Commanders and staffs at all levels are directly involved in the organization of the

information space activity in peacetime, in wartime, in the preparation and execution phases of operations (warfare). Each of these command structures, with regard to their functions and authority, plans the subordinate troop activities linked by a single concept of action in the information space."[132]

To conclude the overview of the doctrine, two distinctive principles are laid out by the authors:
1. the development of an information security system for cyberspace military conflict prevention and,
2. the creation and maintenance of information warfare forces and means in a state of constant readiness to rebuff military and political attacks in the information sphere.

It is particularly interesting because, by "a system," in the understanding of Russian perspectives it is not only a communication and information system, but also a set of tools, measures, concepts, and doctrines for ensuring the achievement of prevention and settlement of conflicts in information space. Such tools include active research and development of techniques and means to address threats coming from information space, as well as having the capability to conduct operations on behalf of Russian strategic interests. Secondly, the Russian armed forces are not only meant to repulse military attacks but also political ones, blending military and non-military means of influence and might. Military assets could be utilized in countering political, nonmilitary threats in the civilian global information sphere.[133]

In an 2013 article written by Russian Armed Forces Army General Valery Gerasimov, the Chief of General Staff emerged in *Voyenno- Promyshlennyy Kuryer*,[134] while not an officially sanctioned doctrine or strategy, offered the perspective of a leading figure of the Russian Armed Forces. The article, "The Value of Science Is in the Foresight: New Challenges Demand Rethinking the Forms and Methods of Carrying out Combat Operations"[135] details the offensive role of psychological and technical cyber activities. Gerasimov states that nonmilitary means of coercion, influence, and pressure may have higher yield in achieving strategic objectives than conventional military tools.

"New information technologies have enabled significant reductions in the spatial, temporal, and informational gaps between forces and control organs."

Gerasimov argues that information and non-military hostilities are already part of the conflict and, as clandestine methods in information confrontation gains prominence, the lines blur between levels of command and control and the hierarchic top-down model.

"Asymmetrical actions have come into wide-spread use, enabling the nullification of an enemy's advantages in armed conflict. Among such actions are the use of special operations

forces and internal opposition to create a permanently operating front through the entire territory of the enemy state, as well as informational actions, devices, and means that are constantly being perfected."

He reaffirms the posture of information weapons having the impact to offset an opponent's conventional military technological advantages.

"Developing a scientific and methodological apparatus for decision-making that takes into account the multifarious character of military groupings (forces) is an important matter. It is necessary to research the integrated capabilities and combined potential of all the component troops and forces of these groupings. The problem here is that existing models of operations and military conduct do not support this. New models are needed."

Conducting information operations also impacts tactical, operational and strategic levels regarding the combat potential of opposing forces. Gerasimov reflects on existing models for decision-making and their vulnerability by advocating for the development of new models reflecting the multifaceted nature of armed conflict where the lines between military and nonmilitary means are blurred. The peace and wartime phases of a conflict overlap; as such, information warfare is an excellent tool for achieving strategic objectives, influencing all three levels (tactical, operational, strategic) of operations. Systems, because they are not accustomed to these new features, are equally vulnerable to them. The adversary's abilities and methods are well understood by Russian military leaders and have a direct impact on their thinking.

> A scornful attitude toward new ideas, to non-standard approaches, to other points of view is unacceptable in military science. And it is even more unacceptable for practitioners to have this attitude toward science.[136]

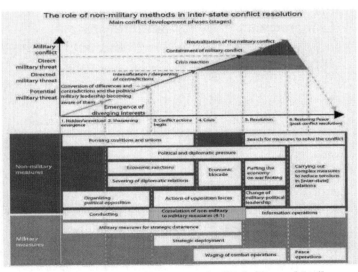

Figure 5: General Valery Gerasimov, the Chief of General Staff emerged in Voyenno- Promyshlennyy Kuryer In McDermott, R. (2014). Gerasimov Unveils Russia's 'Reformed' General Staff. Eurasia. Daily Monitor Volume: 11 Issue: 27. [Online].

MILITARY DOCTRINE OF THE RUSSIAN FEDERATION (2015)

The Doctrine of 2015 again builds upon its predecessor documents, and lists information as an instrument of non-violent nature to protect the interests of Russian Federation and its allies, and that military risks and threats are part of the information space.

The strategy points out external risks such as the use of information and communication technologies in military and political activities against international law, sovereignty, political independence, territorial integrity, international peace, security and global and regional stability. That is the broad scope of areas of impact.

The military doctrine also reiterates and reinforces its emphasis on subversive information operations, aimed at the cultural and moral traditions of the "Motherland."[138] As with the previous military doctrine from 2010, the doctrine from 2015 reiterates the features and specifics of current military conflicts by listing informational measures, this time taking into account the protest potential of the population of the target country.

Among other features described in previous doctrines, the one from 2015 also expands on the new weapons platforms, and as stated in the analysis of the last doctrine, all these platforms are inherently dependent on information and communication technologies.

"...massive use of weapons and military equipment systems, high-precision and hypersonic weapons, means of electronic warfare, weapons based on new physical principles that are comparable to nuclear weapons in terms of effectiveness, information and control systems, as well as drones and autonomous marine vehicles, guided robotic weapons and military equipment;"

The global information sphere, along with airspace, outer space, land, and sea, are domains through which the military and non-military means to develop and sustain pressure on the enemy should be wielded. Combination and coordination through these domains are key challenges in using kinetic effects and connecting them with disruption in the global information sphere to achieve military and nonmilitary objectives.

Informational measures are not only used to strike, disrupt, deceive or just attack the enemy; they also serve as an excellent enabler to conduct analysis, acquire data necessary for decision-making, and to support sensory awareness. Therefore, modern technical means and information technologies are outlined yet again in anticipation of development:

"to assess and forecast the development of the military and political situation at global and regional levels, as well as the state of interstate relations in the military-political field with the

use of modern technical means and information technologies;"

To conclude with the doctrine, unlike the one from 2010 calling for the development of forces and resources for information warfare, the Doctrine of 2015 postulates a need to enhance the capacity and means of information warfare. A distinct advantage is present in going through all of the analyzed strategic documents dealing with national security and defense. It may seem that they are repetitive and the reader might be overwhelmed with threats, thrusts, aims, and objectives. Yet they repeatedly underscore a familiar trend: the commitment to the issues of information warfare, information space, and influence in all of the documents reviewed above shows the degree of prominence of information struggles in the mind of Russian national security and military thinkers. In the strategist's mind, it is of utmost importance to preserve the domestic control of the flow of information, perception of reality and to project purposeful images of reality abroad. The same applies to information superiority vis-à-vis the enemy. In conclusion of this chapter, the ability to gain information superiority is essential to achieve strategic objectives in today's interconnected world. The Russian establishment perceives a technological disparity yet again, so it compelled to control the information sphere by other means. Thus the art of strategic deception and altering decision-making has been reintroduced into the Russian operational art. The ability to conduct strategic deception in cyberspace, the information sphere, has been proven to be one of the key objectives in ensuring national security and defense.

CYBER RELATION TO INFORMATION SUPERIORITY AND REFLEXIVE CONTROL

In this chapter, the emergence of a single information space will be covered, along with how exploitation of this space enables the alteration of the global balance of power. With the expansion of information and communication technologies, the democratization of knowledge, the spread of the internet and other networks for sharing information people have been given unprecedented access to information, which allows a variety of choices and avenues of decision-making.

Two hundred years ago, the primary commodity in interpersonal and international relations was the information itself. The ability to possess and provide information was of significant influence. Knowledgeable people were respected for the information they obtained and were able to use. However, the storage of data has shifted towards digitalized instruments. The amounts of data available are so significant that the commodity is not the information itself, but the ability to navigate through it and choose the right one to serve the purpose in decision-making. The equation from the past, when information was scarce, has shifted to the state where the amount of data is abundant, sometimes monstrous in size. People have started to store information and thus ceased to carry it along. The future includes the ability to analyze the piles of data and to decide what is relevant and vital and truthful. This new paradigm enables confusion, the undermining and spreading of information to either manipulate or to obfuscate the critical information. As described in previous chapters, decision-makers need information in order to conduct conscious decisions, and there are many ways to disrupt that process. If the aim is to minimize or cripple the ability to command and control, there are varieties of techniques for how to achieve that aim.

Although the former Soviet Union was considered a military superpower, it sought alternatives to conventional military power. Years of underfunding during the Cold War and at the beginning of 1990s139 led to the advancement of theoretical and strategically essential concepts such as reflexive control. The imperative was not only to search for alternatives to physical and kinetic capabilities and conventional military might, but to address the changing environment in which the Soviet Union, and later Russia, found itself. The emergence of a single information space threatened the dominance of the state and security apparatus over the flow of information and its availability to the people, ordinary decision makers. It was true not only domestically, where competing information could have severe ramifications for the stability of the regime, but also internationally.

Russian strategists soon realized that – as was the case in the past, on a much smaller scale, when the controller of the situation held the key to achieving its strategic aim – the global information space or unified information sphere was a threat to the worldwide balance of power.140

Whoever controls perception controls reality with definitive impact on the assessment of the importance of information warfare. Russian military theorists and strategists[141] have reached the conclusion that information has become a national or strategic resource[142]. With the dawn of information technologies and the free spread of information in the somewhat anarchic nature of the information sphere, people have unprecedented access to information, a fact which affords everyone – ordinary people to top decision makers – a variety of choices when searching for information and creating their perception of reality.

The informatization of society has penetrated all levels of existence and organizational aspects of the community. Because this includes economic, social regulatory systems and the military itself, this informatization leads to the conclusion that, by influencing the channels, the filters and the content, the controlling subject can achieve information superiority. Information superiority leads to attaining strategic goals without exclusive reliance on conventional power assets. Russian strategists believe that countries that have obtained information superiority will be predisposed to employ military force subsequently.[143] This is likely due to their belief that the controlling subject has control of the perception of reality, and thus is able to form a perceived forcible outcome. Also, military objectives may seem easier to achieve with the support of a complex influence operation. Although this is a position where influence operations are complementary to military activities, the theoretical approach presented by General Gerasimov illustrated the dynamic by stating that the convergence between nonmilitary and military is desirable and should be employed.

Other characteristics of cyberspace which make it suitable for applying information operations for reflexive control are – absent international legislature – fewer legal restraints, less attribution to attacks and less enforceability. With a virtually nonexistent code of conduct combined with the features of the physical world, where hard evidence is obtained quickly in comparison to cyberspace, cyberspace has become a preferred avenue for influence operations – not only because of its amplification features, but also because of the numerous ways to mask the identity and source of the attack. Where there is no crime, there is no prosecution. National legal frameworks differ in dealing with cybercrime, critical information infrastructure protection and cyber terrorism. In the vacancy of binding IHL for cyberspace with states only now trying to create rules of engagement, it is no wonder that cyber-space has become a lawless place ripe for exploitation and information attacks. The examples of individuals indicted for crimes in cyberspace are so few, and too personal – data-theft and illegal contraband-oriented – that pursuing tradecraft relocated to cyberspace is almost too difficult to describe; it is becoming more political with every successful information attack.

Along with the other advantages of cyberspace such as reflexive control and influence operations, low cost of entry pays a significant role. The resource investment requirement here differs from those in conventional military procurement and is an order of magnitude less expensive.

Another facet is the use of proxies to conduct the influence operations, information attacks with hostile code and hostile content on behalf of the controlling subject. These proxies are affiliated indirectly or directly with the control subject. However, a whole market-based economy, reliant on supply and demand exists, which is ironic given the Marx-Leninist roots of the driving theories behind Soviet information operations. In this market, human skills not only can be bought, acquired or rented, but information is for sale, personal data troves useful for creating the profile of an opposing decision maker, stolen information about consumer behavior is available. Also, vulnerabilities and exploits are for sale for malicious code attacks and other computer network operations. Everyone can rent or buy the infrastructure, meaning computational power, for malicious activities. It is a significant shift from the conventional perspective of conducting intelligence operations, where private proxies and unknowing subjects were also utilized, but never on such a scale including all aspects of the action. Even in comparison with the conventional military power, it is hardly imaginable that a control subject would turn to a private citizen to ask for assets of firepower that the citizen is offering on the black market. Nevertheless, it is happening in cyberspace. The privatization and monetization of enabling assets for influencing whole societal systems is a few clicks away.

REFLEXIVE CONTROL:
CYBER-RELATED EXAMPLES

> Computer technology increases the effectiveness
> of reflexive control by offering new methods
> adaptable to the modern era that can
> serve the same ends.[144]

According to one of the Russian military theorists and a writer on information operations, Col. S.A. Komov, several framing methods are described in information operations.[145] He called reflexive control "intellectual methods of information warfare."[146] According to his findings, information superiority is of two types: quantitative and qualitative. In the first case, quantitative information superiority, conventional means of force are utilized to destroy targets when the quantity/availability of information is of importance. In the second case, intellectual methods of warfare are used, whereas reflexive control to manipulate the enemy[147] A military objective is achieved by the combination of both approaches. It is important to reiterate that the targeted vulnerabilities are not necessarily of a technological nature. They could be the aversion to casualties or the enhancement of grievances. Weaknesses can be of technological, doctrinal, organizational or cultural character.[148] However, the technology available today allows for full exploitation.

The elements laid out by Komov and paraphrased by Thomas are:[149]

*"**Distraction**, by creating a real or imaginary threat to one of the enemy's most vital locations (flanks, rear, etc.) during the preparatory stages of combat operations, thereby forcing him to reconsider the wisdom of his decisions to operate along this or that axis;"*

In April 2015, when the international community, including the United States and the European Union, were increasingly worried about the ongoing military campaign of Russia in Ukraine, pressure on the regime in Moscow was increasing. Diverting the international interest and lessening the pressure of potential countermeasures aiming at the administration in Moscow, a threat of higher significance emerged in the arena of international security agenda. Based on the correct perception that populations in the U.S. and Europe are more receptive to the threat of terrorism than a paramilitary operation in Ukraine, an attack was allegedly conducted by the Islamic state, Daesh, as an information operation in cyberspace. Cybercaliphate, a group affiliated as a proxy to Daesh, attacked the French television TV5Monde, affecting it by taking down all 11 channels, taking over social media accounts, website and the infrastructure of TV5Monde itself. The attack, among other vectors, resulted in displaying Daesh's flag. Or at least that was supposed to be the perception, to alter people's interest and to create pressure on political representation, diverting their attention towards more immediate issues such as Daesh. The reality of Daesh attacking Europe would lead to lessening the burden on Russia regarding their military operations in Ukraine. Several expert groups and media outlets attributed the attack to the proxy group APT28[150] aka Fancy Bear (aka Sofacy group, aka Pawn Storm), through means of circumstantial evidence regarding the technical infrastructure used for the attack. Tools, techniques, and procedures assigned to this group served as evidence.[151]

This was one of the first, and best, examples of how a previously military-oriented approach had a civilian application to meet the same ends. By nurturing fear, and creating a real or imaginary threat to the society, it was demonstrably easy to shift the focus of decision makers away from the issue at hand.

*"**Overload**, by frequently sending the enemy a large amount of conflicting information;"*

Overload in the past served as means of providing lots of information to the opposing command structures to limit the decision-making process and to force hastily prepared decisions. Presently, it is still a valid strategy, but technology offers the technical aspect of jamming, done by limiting the command structures to react at all, due to the information processors. The overloading of processors (human, technical, intellectual) with false or conflicting information works well. From a technical aspect, execution of a denial of service that is essential to the command and control structures introduces another type of overload. In August 2014, CyberBerkut allegedly used a DoS[152] attacking the cell phones of the Verkhovna Rada, the Supreme Council of Ukraine, and a DDoS[153] to attack more than 500 hundred web-based information sources.

"Paralysis, by creating the perception of a specific threat to a vital interest or weak spot;"

On Friday, February 28th, 2014, unidentified uniformed personnel overran several key telecommunication nodes of state-owned telecommunication provider Ukrtelecom and physically damaged the infrastructure, including the fiber backbone. On top of that, cell phones were jammed, disrupting communications on the Crimea peninsula. Two days later, attacks on the power lines and communication infrastructure of the Sevastopol Navy command were conducted to sever communication and disrupt command and control. Also, the head of the Ukrainian security service (SBU announced on March 4th 2014 that equipment to block communication among parliamentarians and other decision-makers efficiently crippled the connection between them and the ability to coordinate in the time of national security crisis. However, before the attack on the cell phones, landlines were also affected, pushing the decision makers to turn to cell phone coverage that has been compromised by the installation of unknown devices at the Ukrtelecom nodes. It is a classic example of Komov's combination of quantitative and qualitative information superiority.[154] Compromising communication network either to deny any information exchange that is instrumental in understanding the reality on the ground, and employment of qualitative measures when compromising the other means of communication, effectively left the decision makers in doubt about whether the infrastructure could be used or not.

"Exhaustion, by compelling the enemy to carry out useless operations, thereby entering combat with reduced resources;"

Another example is to exhaust the capacities devoted to cybersecurity on incidents that have national significance or criminal implications in order to drain the potential network defenses. A government cannot cease to investigate cybercrime or cybersecurity incidents, but an unprecedented rise in the number of events and cyber-criminal activities can efficiently exhaust the government's ability to respond. Based on the testimony of a high ranking official from Kiev dealing with cybercrime, prior to the Russian campaign in Crimea and Eastern Ukraine, Russians cooperated with Ukrainians regarding cybercrime on a daily basis. This cooperation inadvertently led to understanding the structure, ranks, individuals and responsibilities and decision-making algorithm, one not unlike the Russian system. However, once the campaign started, some of the assets the law enforcement and cybercrime units had in Crimea stopped communicating with headquarters, leaving only a few people with this skillset available in the region. A significant spike in cybercrime cases led to the exhaustion of the cybersecurity and cybercrime countering capacities. Silence from former partners in Russia was deafening. In one example mentioned by the ranking official, Ukrainian law enforcement authorities were cooperating with their Russian counterparts on a case involving infrastructure located in Moscow and being used for illegal, malicious activity. However, the servers physically localized by the operation of law enforcement agencies disappeared after the military campaign to occupy Crimea commenced. The servers eventually turned up again in a remote location and were used to conduct cyber-attacks against Ukraine. A high-ranking law enforcement official from Ukraine provided the information during a closed-door interview.[155]

Exhaustion was mainly intended for military and troops. However, with the lines blurred, the skill set of cyber-oriented law enforcement personnel and military-aged men, called to arms to participate in military operations, also had the effect of exhaustion of civilian assets called to arms.

*"**Deception**, by forcing the enemy to reallocate forces to a threatened region during the preparatory stages of combat operations;"*

Deception in the case of hostilities and military operations is a well-known concept throughout history. Electronic warfare has been used countless times to jam communications, present false evidence of a weak spot to induce the enemy to allocate forces, to create a false assumption in the decision-making process that a particular location is under threat, which then leads to a weakening flank or rear of the troops. What is new is the ability to broaden the scope of deception campaigns using technology as propagators of deceptive messaging. As commanders have to rely on information from the field, jammed communications will eventually turn them to other sources to enrich their information resources. This is the time when using open source or less-protected encrypted communication channels come into use. It is easier to jam, or block communication, either technically or by creating the perception that they have been compromised, than to turn the commanders and decision makers to use less secure or legitimate, but still working, sources and channels. The injection of false messaging, combined with overload creates a distortion in the decision-making algorithm, either by providing deceptive information leading to faulty conclusions, or by mixing false messaging with an overload element on the intellectual level to both strain the timeframe for the decision and create a vague situation to the advantage of the control object. By efficiently harming the decision-making process in the command and control structures, the controlling entity achieves information superiority, which leads to informational control over the battlefield. Given the concept of reflexive control and having identified the mechanisms of command and control, the controlling entity can create deception elements that lead to predetermined decisions made by the controlled subject. Examples abound of such activities and false reporting: word of units spotted in a location that should have been cleared of enemy forces (in this case, Russian special forces); fake news coming from Crimea about the size, strength or combat readiness and equipment that is available. The confusion has in instances led to Ukrainian forces being lured into ambushes, due to fake radio transmissions aired by the opposing forces, due to a false perception of the situation on the ground, and on account of jammed or hacked cell phones used to communicate between units and battalions. Deception is nothing new, but the technological advancements in past few decades have opened a new dimension of its advantages.

*"**Division**, by convincing the enemy that he must operate in opposition to coalition interests;"*

This element, repurposed for civilian decision-making, has a significant potential to alienate the public and cabinet coalition parties. It serves to undermine support for decisions and actions by sowing distrust between the coalition partners and exploiting historical grievances within society. The division is again a well know concept best described as *Divide et Impero* in Latin allowing the controlling subject to undermine the ability of the control object to act in concert once the cohesion of the deciding entity or public support is shaken. One way of alienating partners and sowing distrust in the society is to claim false information for the purpose of angering the respective parties, either by targeting the electorate or magnifying historical grievances – or, as in the case below, harming the relationship between NATO members Estonia and Ukraine.

On April 26th, 2015 an article on a pro-Russian news outlet RT[156] presented the readers with an accusation, that the NATO Cooperative Cyber Defence Centre of Excellence[157] based in Tallinn Estonia was involved in the establishment and technical assistance to the Mirotvorce webpage. The allegations were brought up by the RT news channel, creating the perception that the CCD CoE (which is an influential intellectual and technical source of cyber security and cyber defense knowledge), was involved with a Ukrainian terror site, insinuating that NATO is directly responsible for running the site. The site served, among other purposes, as an aggregator of so-called enemies of the state, opposing the authorities in Kiev providing detailed information about people considered as enemies of the state. The report by RT was intended to influence its audience towards the perception that NATO is supporting the website and is thus targeting journalists, activists and members of parliament not in favor of official policies of the government. The inclusion of NATO and the Center in Estonia was meant to spur unrest among people, that international meddling of in internal affairs is happening and supported by NATO. Also, the labeling of Kiev junta and accusations of outright fascism were present in the article. After official protests against the report, RT issued a correction on the page of the story, but left the story in place and chose evasive language in the revision.

NATO trace 'found' behind witch-hunt website in Ukraine

Figure 6: RT, (2015). NATO trace 'found' behind witch hunt website in Ukraine. [Online]. Available at: https://www.rt.com/news/253117-nato-ukraine-terrorsite/

CORRECTION: After a thorough revision of the published story we have found out that the domain name Psb4uKr.ninja, which was claimed to be registered by NATO CCDCOE, is a mirror of the original Mirotvorec website. We admit that we cannot confirm any credible link between NATO's Cooperative Cyber Defense Centre of Excellence and Mirotvorec.

Trends
Ukraine turmoil

NATO's Cooperative Cyber Defence Centre of Excellence – CCD-COE has reportedly been exposed as providing technical support for Mirotvorec, a website of Ukrainian nationalists running 'enemies of the state' database.

The information available at DomainTools, *"the leader in domain name, DNS and Internet OSINT-based Cyber Threat Intelligence and cybercrime forensics products and data"*, is decisive: the registrant of the Mirotvorec website is 'NATO CCD-COE' and its employee 'Oxana Tinko', operating from Estonia's capital Tallinn. The address of the registrant coincides with the address of CCD-COE: Filtri tee 12, Tallinn 10132, Estonia.

📷 Screenshot from http://whois.domaintools.com

In March 2014, there were media reports that 16 employees of CCD-COE were detached to Kiev to provide cyber security support to Ukraine.

The hacktivist group CyberBerkut, which has been opposing Kiev authorities from the very beginning of the unrest in Ukraine in 2013, claimed responsibility for taking down three NATO websites in a series of DDoS attacks a year ago. CyberBerkut claimed it brought down NATO's main website (nato.int), as well as the sites of the alliance's CCD-COE cyber defense center and NATO's Parliamentary Assembly.

The hacktivists claimed that they are countering the action of the so-called 'Tallinn cyber center' or NATO Cooperative Cyber Defense Centre of Excellence, which has been hired by the *"Kiev junta"* to carry out *"propaganda among the Ukrainian population through the media and social networking."*

Figure 7: ibid.

*"**Pacification**, by leading the enemy to believe that pre-planned operational training is occurring rather than offensive preparations, thus reducing his vigilance;"*

In conventional military campaigns, making the enemy believe that military exercise is not a preparatory phase of an imminent offensive campaign is essential. Reducing the vigilance of enemy troops and commanders is key for exploiting the moment of surprise. However, pacification can also serve a psychological purpose. Inhabitants of war-torn regions in Ukraine had been received messages via their cellphones; even radio broadcasts had been reported exercising psychological pressure on the recipients and audience. Such activities aim to pacify the resolve of the troops, or to threaten and coerce the general population. Among some of the messages were ones that targeted the relatives of soldiers of the then-Ukrainian military, informing them that either their child died or will be inevitably killed. Again, suppressing the resolve, undermining the will to fight and disseminating antagonistic feelings towards a military operation itself is not a new technique. However, with the spread of mobile phones and related technologies, the term "pinpoint propaganda" has reached its peaks. In the past, geographic factors, and linguistics and technical limits of pacification messaging defined the dissemination and broadcasting. Leaflets were dropped indiscriminately over areas of perceived impact zones. Today, courtesy of mobile technologies, the ability to pinpoint exact locations, choose the time of sending messages, and having them immediately impact recipients along with other preplanned activities, is of great value. The technology, however, enables even more. Cell tower simulators send messages carrying threats and disinformation. The simulator pretends to be part of the cellular, GSM network serving as a surveillance tool by law enforcement, or intelligence agencies to monitor the whereabouts and get hold of actionable intelligence from the location of the cell phones. In the cases of Crimea and Eastern Ukraine, these cell tower simulators are called IMSI catchers. They serve as a "man-in-the-middle" type of tool, where they interfere and acquire information between the cellphone and the real tower of the service provider. IMSI stands for International Mobile Subscriber Identity, which enables the identification of specific mobile devices in the network, and to target them. The contents of the message stating that the soldiers should leave or die was sent to the targeted cellphones in waves over time since the eruption of hostilities in 2014.

Some messages are also disguised to look as though originating from other units, containing information of desertion among commanders in the area. Messages seemingly coming from fellow soldiers have also been recorded. The Ukrainian national authorities acknowledged this type of information warfare through the statements of Col. Serheyi Demedyuk, then Head of the National Cyber Police Unit, for media.[160] The impact of receiving the tailor-made messages for specific target groups is present, as controlled objects tend to pay more attention to a message arriving at their phone than to a leaflet falling from the sky. In 2015, an article in the *Russian Military Review*[161] stated that the LEER-3, a truck-mounted electronic warfare system has a drone with the ability to simulate cell tower services, serving as an IMSI catcher with the capacity to hijack over 2000 cellphones in a 6 km wide area. Informnapalm, a volunteer group of open source intelligence analysts elaborated on the claims, with an article about the LEER-3 platform, operated by Russian Armed forces, with pictures and videos locating the electronic warfare system in the occupied Donetsk.[162]

*"**Deterrence**, by creating the perception of insurmountable superiority;"*

Deterrence can be achieved by several means, one of them by promoting the perception of invincible superiority. In the face of Russian aggression in Ukraine, the U.S. Navy dispatched an Arleigh Burke-class destroyer equipped with the Aegis system, to the Black Sea. The ship, the USS *Donald Cook*, was in international waters. The Aegis combat system is an advanced warfare platform and part of the ballistic missile defense system. It provides intelligence for decision-making in combat, underwater, surface and air/space monitoring and targeting of threats, and coordinates precision strikes with multiple weapon systems across different platforms such as surface vessels, fighter jets, and ground weapons platform. Therefore, the ability to showcase the debilitation of a system of such importance is an exciting case of perception of technological superiority, and serves as a potential deterrence. To promote such supremacy without engaging in hostilities and direct combat activities, a demonstration of electronic warfare can be a suitable option. Electronic Warfare (EW) is a powerful tool for suppressing communication, jamming, spoofing or otherwise disrupting the intended functions of electronic systems. Taking down the Aegis system with an EW system would be a powerful signal to the domestic audience, as well as to the international community. In fact, this avenue of thought was followed already in 2014.

In April 2014, the USS *Donald Cook* was allegedly jammed, deeming its advanced AEGIS system inoperable and crippling the destroyer's capacity to fight altogether. The reported incident described the Arleigh Burke class vessel sitting on the sea in a blackout after a Russian SU-24 overflew the ship and used the Khibiny EW system. The SU-24, which is an aircraft that can be mounted with anti-vessel rockets, did several flybys to demonstrate its superiority. It resulted, according to the reporting, in the immediate recall of the *Donald Cook*, and had a profound psychological impact on the crew. Reports of this incident were prevalent in pro-Russian news outlets, online, print, radio. It was reported with more skepticism by western tabloids, but made its way into the arena of alternative, bogus websites. The story was first published in 2014 and resurfaced in 2017. The incapacitation would be a perfect example of demonstrating superiority, thus enhancing deterrence through a technological edge. Although the incident never happened, it was widely reported, covered, and had the intended impact,[163] despite being repeatedly debunked.

To properly elaborate on the creation of disinformation, committing deception against the domestic audience and having the international community purposefully misguided in an event, which never happened, this author examines the incident from the timeline perspective. The analysis is based on the findings of other authors that support, and in higher detail describes, the incident by the Digital Forensic Research Lab.[164]

The Arleigh Burke-class destroyer USS *Donald Cook* entered the Black Sea on April 10th, 2014. On April 12th, a Russian military aircraft SU-24, NATO codenamed *Fencer*, conducted a flyby and overflew the vessel several times. On April 14th the Pentagon issued a statement[165] denouncing the behavior of the pilot commenting on the close range flyby and not responding to the vessel, calling it a provocative action. April 17th was the birthday of perhaps the clumsiest parody turned disinformation example the author has seen. On the portal Fondsk.ru the article first appeared on how the Russian SU-24 debilitated the Aegis-equipped vessel, and then reposted on several social accounts. The wording of this article plays a role in subsequent iterations of this story. The original article, however, was a satiric piece in the "opinion" part of the website, meant to mock the U.S. Navy. Two days later, the same article appeared on a different website, freepress.ru, in the "humor" section.[166] The article suggested that an advanced EW system called Khibiny, mounted on the SU-24, was responsible for taking down all electronic equipment aboard the ship. Interestingly, the opinion piece was written as a letter from the perspective of one of the *Donald Cook* crew members. At this moment, the story about the EW deployment was posted no more than twice, and as a reposted bogus story on social media. Nevertheless, it caught the attention of the state-run propaganda machine, Sputnik (a Russian government sanctioned and financed propaganda news channel), which on April 21st 2014 in its German mutation[167] reported that an SU-24 had crippled a U.S. destroyer using an advanced EW weapon, sourcing the story from Russian media and bloggers.

Interestingly, the story with USS *Donald Cook* in a central role had another path on the novorus.info portal, connected with the separatists in eastern Ukraine. The story claimed that 27 crew members resigned immediately after the flyby of the SU-24. Using a classic tactic, the report cited a Reuters story which, in fact, did not state anything about the resignation. The purpose was to confuse the audience by introducing a respected source, thereby legitimizing the novorus.info story, which had no relation to the EW incident whatsoever.

Another burst of disinformation, through a different channel, appeared on April 30th in the Rossiyskaya Gazeta, the official Russian government-run news outlet for publishing news, new legislation and decrees. This media outlet is under the full control of the Russian government.[168] The article reported with an underlying mockery, that the Aegis was shut down by the Khibiny system, leaving the vessel inoperable for combat operations, claimed that crew members had resigned, and that the ship rushed to port in Romania to "put its nerves together." By now the incident of the close flyby was widely cited and reported, and served as a frame for introducing the disinformation piece about the electronic warfare attack and the resignation of troops, citing sources such as the Pentagon press release and Reuters reporting confirming the allegations. Not many readers, however, went back to verify reverse sourcing. On September 13th, the obscure web portal Voltaire.net picked up the story of Rossiyskaya Gazeta, and translated it into its Spanish mutation. Voltaire.net, a portal for used spreading conspiracy theories and supporting Iranian and Syrian governments, was founded in its current form in Lebanon, and claims registration in Hong Kong.[169] The portal propagates conspiracy theories and anti-Western propaganda through various channels such as websites, online TV, and RSS feeds. The article – initially translated from Russian to Spanish – was subsequently translated into French, Polish, Italian, Arabic, German and Portuguese within weeks. The English version was published in November 2014. See below:

Figure 8: Voltaire Network, (2016). About Voltaire Network.
Available at: http://www.voltairenet.org/article185860.html

Figure 9: ibid.

Voltaire TV

Voltaire Network TV (YouTube)

Facebook

- Français : facebook.com/reseauvoltaire
- Español : facebook.com/voltairenetespanol
- English : facebook.com/voltairenetwork
- Deutsch : facebook.com/voltairenetzwerk

Twitter

- Français : twitter.com/reseauvoltaire
- Español : twitter.com/redvoltaire_es
- English : twitter.com/voltairenet_en
- Deutsch : twitter.com/voltairenet_de

RSS feeds

- عربي
- Čeština
- Deutsch
- Ελληνική
- English
- Español
- فارسى
- Suomi
- Français
- Italiano
- Nederlands
- Norsk
- Polski
- Português
- Română
- Русский
- Türkçe
- 中文

Figure 10: Available at:
http://www.voltairenet.org/article185860.
html

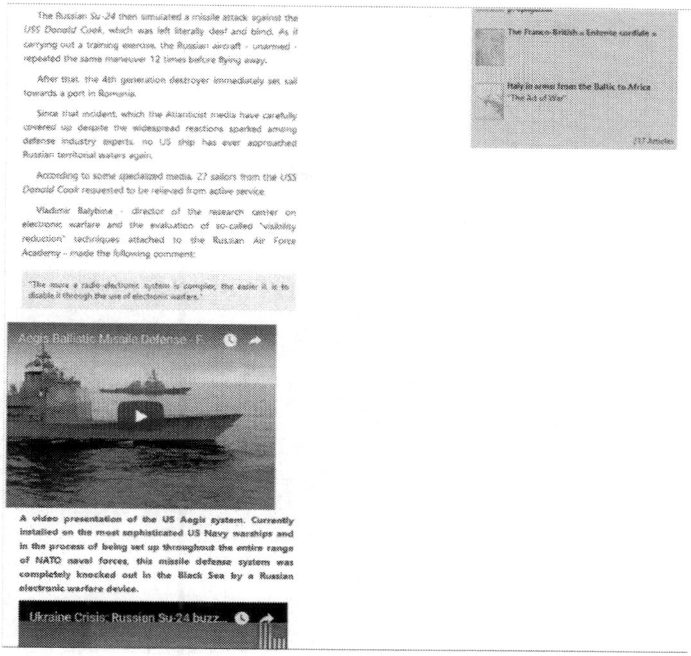
The Russian Su-24 then simulated a missile attack against the *USS Donald Cook*, which was left literally deaf and blind. As if carrying out a training exercise, the Russian aircraft - unarmed - repeated the same maneuver 12 times before flying away.

After that, the 4th generation destroyer immediately set sail towards a port in Romania.

Since that incident, which the Atlanticist media have carefully covered up despite the widespread reactions sparked among defense industry experts, no US ship has ever approached Russian territorial waters again.

According to some specialized media, 27 sailors from the *USS Donald Cook* requested to be relieved from active service.

Vladimir Balybine - director of the research center on electronic warfare and the evaluation of so-called "visibility reduction" techniques attached to the Russian Air Force Academy - made the following comment:

"The more a radio-electronic system is complex, the easier it is to disable it through the use of electronic warfare."

Aegis Ballistic Missile Defense - F...

A video presentation of the US Aegis system. Currently installed on the most sophisticated US Navy warships and in the process of being set up throughout the entire range of NATO naval forces, this missile defense system was completely knocked out in the Black Sea by a Russian electronic warfare device.

Ukraine Crisis: Russian Su-24 buzz...

The Franco-British « Entente cordiale »

Italy in arms: from the Baltic to Africa
"The Art of War"

217 Articles

Figure 11: Voltaire Network, (2016). About Voltaire Network. *[Online]. Available at: http://www.voltairenet.org/article185860.html*

After publication of the article on Voltaire.net in its English version, alternative and outright obscure media outlets picked it up.[174] Portals such as Infowars:[175]

RUSSIANS DISABLE U.S. GUIDED MISSILE DESTROYER

Incident skipped over by the Pentagon-friendly corporate media

Kurt Nimmo | Infowars.com - NOVEMBER 13, 2014   💬 1 Comment

IMAGE CREDITS: MCCA MICHAEL | BUNDESARCHIEV, WIKIPEDIA COMMONS.

Figure 12: Info Wars, (2014). Russians Disable U.S. Guided Missile Destroyer. [Online]. Available at: https://www.infowars.com/russians-disable-u-s-guided-missile-destroyer/

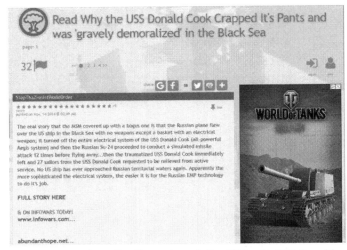

Figure 13: Above Top Secret, (2014). Read Why the USS Donald Cook Crapped It's Pants and was 'gravely demoralized' in the Black Sea. [Online]. Available at: http://www.abovetopsecret.com/forum/thread1042109/pg1

By late November 2014, this might have been just an unsuccessful attempt by the Russian establishment to use a parodic story for the information campaign, employing media outlets such as Sputnik and Rossiyskaya Gazeta, while the story ended up only on the conspiracy side of the news.

On March 26th, 2015 the manufacturer of the Khibiny system, KRET – Concern Radio-electronic Technologies,[178] – published an official article promoting its products, stating that the "famous April attack in the Black sea on USS *Donald Cook* by SU-24 bomber jet allegedly using Khibiny complex is nothing but a newspaper hoax."[179] The manufacturer of the EW system stated that the Khibiny system could not be mounted on SU-24.

In 2017, almost exactly three years after the first reports of the incident that never happened, the story was elevated to again message and boast about Russian technological supremacy over the U.S. Navy using an electronic warfare system. Despite being debunked several times, including by the manufacturer of the weapon system, on April 15th, 2017 it nevertheless resurfaced. The Russian state-controlled TV Rossiya-1 broadcasted a story about the advances of Russian electronic warfare progress. The message of the report was that Russia possesses the capability to defend itself and take out enemy military equipment – whole platforms without a single shot. This is due to the ingenuity of the Russian military-industrial complex. The TV channel provided the incident of USS *Donald Cook* from 2014 and its encounter with the SU-24 as a proof of the effectiveness of the Khibiny system. The report on TV used the previously debunked social media post and its source, fondsk.ru.[180] In fact, the state-controlled TV channel informed its audience about the incident, citing the fondsk.ru parody article word for word, accompanying it with seriously presented background information and stating this event as a fact of insurmountable superiority over the most advanced targeting system of the U.S. military. The *Vesti* news program used many techniques to enhance the reliability of the report, such as mixing the fabricated story with relevant facts, with camera shots from the file showing the Russian military using EW, and camera shots of a destroyer class vessel with a commentary telling the audience how the mighty Aegis was defeated. Also purposely inserted was the statement of an authoritative figure, actually from the U.S. side, a military representative, General Frank Gorenc, former US Air Force commander in Europe, claiming that the "Russian electronic weapons completely paralyze the functioning of American electronic equipment installed on missiles, aircraft, and ships."[181]

Electronic Warfare: How to Neutralize the Enemy Without a Single Shot

Figure 14: Vesti News, (2017). Electronic Warfare: How to Neutralize the Enemy Without a Single-Shot. [Online] Available at: https://www.youtube.com/watch?v=vI4uS307ydk&feature=youtu.be

The report is available on Youtube streaming service with English subtitles.[183]

The report from the established Russian propaganda channel went international within days. The professionalism of the foreign media was defined by how it reported the news about the SU-24 knocked out the Aegis-equipped vessel. UK tabloids such as the *Sun* and *Daily Express* joined the conspiracy websites in deliberating about the incident, with varying grades of skepticism.[184] The news was also picked up by US mainstream media outlets, such as Fox News, and later examined by the *Washington Post* and *New York Times*. The latter ran an account on how a bogus story showed up in the information sphere, and how Fox News somewhat propagated it.[185]

Figure 15: FOX NEWS, (2017). Russia claims it can wipe out US Navy with single 'electronic bomb'. Web Archive. [Online]. Available at: https://web.archive.org/web/20170517 23954/http://www.foxnews.com/world/2017/04/19/russia-claims-it-can-wipe-out-us-navy-with-single-electronic-bomb.html

The issue was not the fact that several western media outlets printed – with or without skepticism – an article from a source that might have been outright propaganda. The issue was the delivery of coverage of the incident on a silver platter for further dissemination and legitimization by alternative and conspiracy communication channels used for spreading disinformation. Channels, linguistics, reverse citations of never reported events, obfuscation of truthful statements, video recording manipulated and streamed as a backing source, official source materials accompanied by authoritative expert opinions all play into the psychological and cognitive processing of information. Dissemination at light speed across the globe, in various language mutations with mutual support between sources – all these techniques provide for cross legitimization, gradual elevation from bizarre websites or streaming accounts to mass media to the mainstream.

The technique of cross legitimization is not new to the digital age. For example, disinformation regarding the origins of AIDS, carefully analyzed in the U.S. Department of State Report on Soviet Influence Activities[187] showed how several witting agents of influence were employed – agents such as quasi-experts, third world newspapers serving as boosters of the story played their role – and how an interesting article was taken over by Soviet media outlets and the story finally reached the Western media outlets.

The present occurrence of disinformation fabricated, enriched and propagated by government-backed pro-Russian sources, is not among the stellar examples of disinformation. The incompetence and clumsiness that allowed debunking, and the decision to revive an already debunked story three years after initial presentation, shows a lack of imagination and craft.

A Swedish study on influence through public diplomacy and active measures noted:

> Like soviet propaganda, Russian public diplomacy today can also be wildly inconsistent. The West is portrayed as weak, but at the same a near existential threat to Russia. Europe is described as xenophobic towards refugees, and foolish for allowing so many of them to seek asylum. The point

however might not be to present the target groups with a coherent alternative narrative. Tools such as Sputnik can also serve the purpose to spread confusion and encourage disunity. Furthermore, the Swedish language Sputnik relied, with few exceptions, on rewrites of already existing news stories from established media outlets.[188]

Nevertheless, available disinformation examples provide us with insight into what tools the Russian establishment has at its disposal and the way it operates in promoting disinformation. The channels, methods, the versatility of sources and propagation based on modern technology and psychological impacts help to shape the perception and distribution of disinformation. Other instances may not be that easy to uncover.

"Provocation, by forcing an opponent into taking action advantageous to your side;"

Provocation is an underlying element in many intelligence activities in general. It helps to shatter the state of the target subjects, create new circumstances, and apply pressure to achieve the desired impact. The provocation can be of various forms, such as direct and indirect.

Achieving indirect incitement involves a group of subjects that are close or related to the subject, such as co-workers or family. Direct provocation is when the activity consisting of provocative behavior is directly targeting the subject; blackmail, for example. Unpleasant information harms the subject's position among his peers, or psychological pressure is applied by exploiting known weaknesses or discovered vulnerabilities. The psychological impact and pressure pays the central role in provocation, as not only the content of the provocation plays its part, but also the delivery method (such as close aides, family, trusted peers or channels) is essential and can achieve good results. For example, if a provocative behavior is expressed by someone with whom the subject is unfamiliar it can be dismissed. However, if the provocative behavior or activity is carried out by someone trusted and respected, then the effect has a different gravity, although the character of the provocation was the same.

Therefore, channels and tools used for the provocations have the same importance as the provocation itself. The methods of delivery and dissemination in the digital era make it possible to target large numbers of subjects, or attack individuals through several vectors that are seemingly unrelated. This can be achieved by using human assets, whose provocation activities are supported by leaked information to the press, or pressure over social media and online media outlets.

On October 24th 2014, CyberBerkut — a group of organized pro-Russian hacktivists hacking on behalf of the Russian Federation against the central government in Ukraine, NATO and Western corporations among others — performed a hack of digital billboards in downtown Kiev, the capital of Ukraine. The timing and choice of channel was significant.

President Poroshenko declared the elections to take place on October 26th. The polls were supposed to change the powers in the Verkhovna Rada, and were early, as Poroshenko wanted to consolidate political forces and oust representatives of former President Yanukovich.

Because of the ongoing military campaign and annexation of Crimea, several regions of Ukraine were exempted from the elections. CyberBerkut, fighting against the central government in Kiev, took that as a pretext and one of the underlying messages of the cyberattack, which constituted of hostile code and hostile content. The aim was to alienate the general population against the political elites in Kiev, showing their distance from the ordinary Ukrainians and the soldiers. The attack used several billboards during traffic peak hours to expose as many people to their provocative content as possible. The hacked billboards showed pictures of political leaders that were anti-Russian, along with inflammatory text and gory pictures from the frontlines and mass graves, illustrating for the electorate that these politicians are fighting against their own people.[189] The billboards displayed this for several hours.[190]

'Cyberberkut' hacked Kyiv billboards

Figure 16: CyberBerkut (2014). CyberBerkut hacked Kiev digital billboards. [Online]. Available from: https://cyberberkut.org/en/olden/index2.php

Static billboard space is commonly used to promote political agendas. Nevertheless, using a digital billboard enabled the attackers to showcase video footage, connecting it with the pictures of politicians and public personae to trigger a change in perception. The displayed personalities were tied to the pictures of gore. Psychologically, it may have impacted the electorate on an emotional level; that was the aim of the hackers. Hacking digital billboards became a stunt performed across several countries for various purposes, so much so that researchers gave a presentation on this type of attack at the DefCon[192] conference.[193] The use of digital marketing tools, such as billboards, to impact large quantities of subjects can be expected to be on the rise.

This type of provocation, aimed at an electorate shortly before elections, was not only a provocation in itself – it is clear that the intent was to change perception, and in some cases, the election results.

"**Suggestion**, by offering information that affects the enemy legally, morally, ideologically, or in other areas;"

The suggestion can be made by providing the opposing side forgeries to undermine resolve, or provide the opponent with information or perception unfavorable to his objectives. One of the prime examples is the alleged hack[194] conducted by CyberBerkut, of Ukrainian Colonel Pushenko who worked at the Ministry of Defense. The suggestion lies in fact that the colonel's email was allegedly hacked and forged documents regarding numbers of deserted Ukrainian soldiers were leaked to the media via the Internet. This had the aim to undermine the morale of the troops and the society showing the high numbers of deserters and also the amounts of lost military equipment taken over by the militias in Crimea and Eastern Ukraine. CyberBerkut describes it in detail:[195]

> The colonel also reports huge losses of military vehicles. Most of them were captured by voluntary military forces of Novorossia. Only in a period of one month from June 20 to July 20 the punishers "gifted" to the voluntary military forces a great number of armory: tanks T-64 – 25 units; infantry fighting vehicles (BMP) – 19 units; armored personnel carriers (BTR) – 11 units; self-propelled guns (SAU) 2S1 Gvozdika (M1974 NATO index) – 11 units; multiple

rocket launchers (RSZO) BM-21 Grad (M1964 NATO index) – 12 units; howitzers D-30 – 5 units; 82 mm. caliber mortars – 16 units; anti-aircraft mounts ZU-23-2 – 2 units; artillery tractors (AT) – 5 units.

The second example of suggestion, albeit very awkward, is the video[196] released by pro-Russian separatists (Lugansk People's Republic) claiming discovery of U.S. Army FIM-92 Stingers[197] at the Lugansk airport, Ukraine.[198] Footage of the video was supposed to have provided evidence of the U.S. Army providing illegal military supplies to the Ukrainian military. The footage showed wooden cases with U.S. Army inscriptions, and FIM-92 Stingers allegedly provided to the Ukrainians by the U.S. Department of Defense. However, the depicted stingers in the video bear the same spelling mistakes as the Stingers from the video game Battlefield 3, raising questions. The suggestion here was twofold: first, that the United States was secretly providing lethal military equipment to the Ukrainian military, and second, that these weapons were to be used against civilians in the area.

"**Pressure**, by offering information that discredits the government in the eyes of its population."[199]

On October 25th 2014, when the elections to the Parliament of Ukraine, Verkhovnaya Rada, took place, CyberBerkut announced that it had gained control over the systems and infrastructure of the

Ukrainian Central Election.²⁰⁰ Allegedly, the electronic vote counting system was attacked, and the Central Election Committee website was down. This pressure was to erode the trust of the population in the state authorities to provide and to secure fundamental citizen rights such as voting. The significance was not in the technical side of the attack, but rather in the psychological impact on the population.

To provide readers the extent of cyber operations and activities in the information sphere, the author prepared a date/element framework called DOPES. The DOPES framework stands for Distraction, Deception, Division, Deterrence, Overload, Paralysis, Pacification, Provocation, Pressure, Exhaustion, and Suggestion. It is based on the framing methods and elements proposed by Komov. It is used here to detail incidents between February 2014 and June 2015 related to the annexation of Crimea and the military conflict in Eastern Ukraine. This framework is intended to help orient readers to the role of seemingly isolated cybersecurity incidents from a broader perspective of information warfare. This list of elements of information warfare is not definitive, nor do the features stand alone. They often overlap, regarding techniques used, objectives targeted, and impacts intended or achieved. Although it may seem limited in scope, the author picked the following list of incidents that occurred and were well documented in the public domain and the records of the National Cyber Security Center during the stated period.

The list of incidents is ordered in a way that presents the date of the event, the DOPES elements it represents and a short description of the incident. It is necessary to note that the list is not exhaustive, as many episodes were either unreported, remain classified or were not logged by the sources used by the author. Thus, the purpose of this table is illustrative.

CYBER-ATTACKS AND INFOCYBER-CYBER-ATTACKS AND INFORMATION OPERATIONS RELATED TO THE CRIMEA AND EAST UKRAINE CAMPAIGN

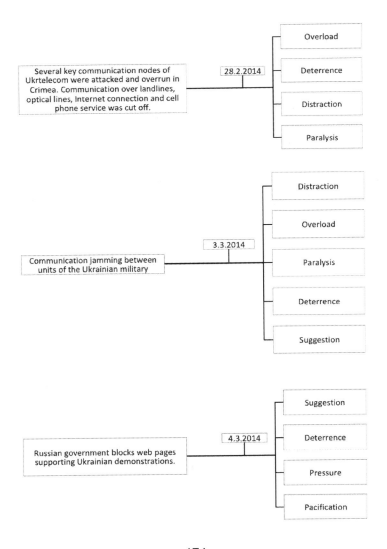

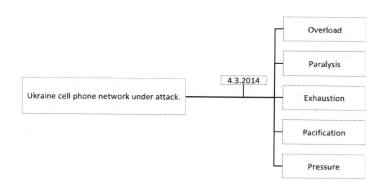

Ukraine cell phone network under attack. — 4.3.2014
- Overload
- Paralysis
- Exhaustion
- Pacification
- Pressure

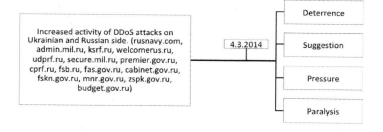

Increased activity of DDoS attacks on Ukrainian and Russian side. (rusnavy.com, admin.mil.ru, ksrf.ru, welcomerus.ru, udprf.ru, secure.mil.ru, premier.gov.ru, cprf.ru, fsb.ru, fas.gov.ru, cabinet.gov.ru, fskn.gov.ru, mnr.gov.ru, zspk.gov.ru, budget.gov.ru) — 4.3.2014
- Deterrence
- Suggestion
- Pressure
- Paralysis

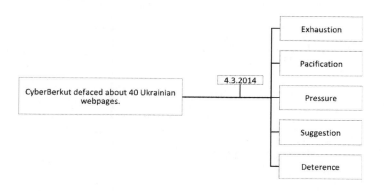

CyberBerkut defaced about 40 Ukrainian webpages. — 4.3.2014
- Exhaustion
- Pacification
- Pressure
- Suggestion
- Deterence

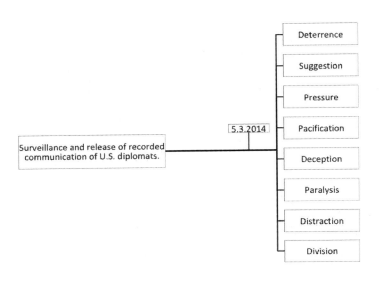

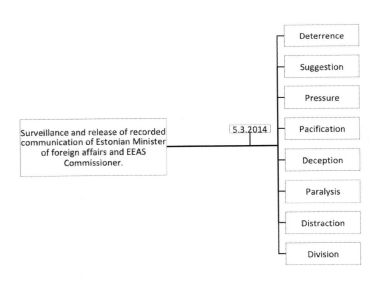

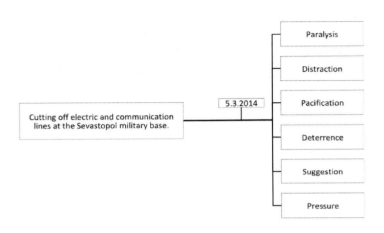

Cutting off electric and communication lines at the Sevastopol military base. — 5.3.2014

- Paralysis
- Distraction
- Pacification
- Deterrence
- Suggestion
- Pressure

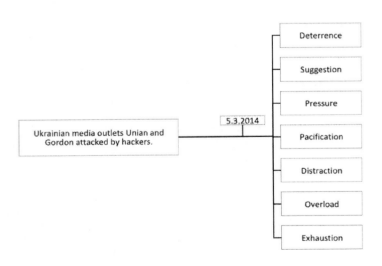

Ukrainian media outlets Unian and Gordon attacked by hackers. — 5.3.2014

- Deterrence
- Suggestion
- Pressure
- Pacification
- Distraction
- Overload
- Exhaustion

Government computers of Russia, France, Lithuania, Latvia, U.S., Belgium, Kazakhstan, Georgia, Italy, Poland, Germany and Jordan were attacked by spyware Turla.

7.3.2014

Distraction

Pressure

Suggestion

Deterrence

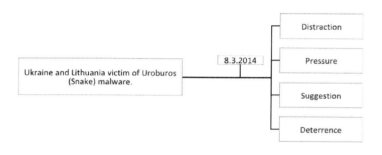

Ukraine and Lithuania victim of Uroburos (Snake) malware.

8.3.2014

Distraction

Pressure

Suggestion

Deterrence

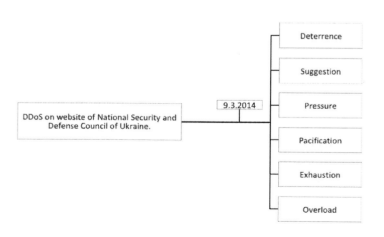

DDoS on website of National Security and Defense Council of Ukraine.

9.3.2014

Deterrence

Suggestion

Pressure

Pacification

Exhaustion

Overload

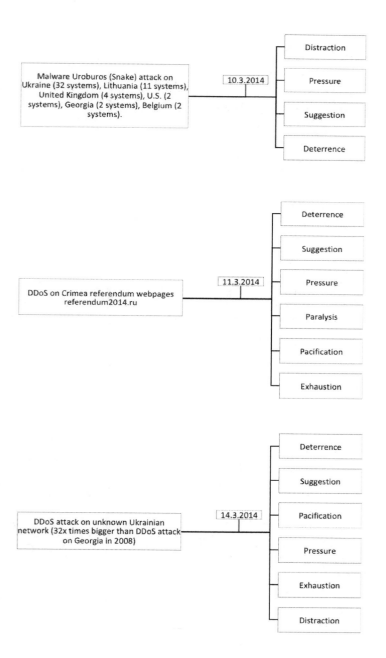

Malware Uroburos (Snake) attack on Ukraine (32 systems), Lithuania (11 systems), United Kingdom (4 systems), U.S. (2 systems), Georgia (2 systems), Belgium (2 systems).

10.3.2014

- Distraction
- Pressure
- Suggestion
- Deterrence

DDoS on Crimea referendum webpages referendum2014.ru

11.3.2014

- Deterrence
- Suggestion
- Pressure
- Paralysis
- Pacification
- Exhaustion

DDoS attack on unknown Ukrainian network (32x times bigger than DDoS attack on Georgia in 2008)

14.3.2014

- Deterrence
- Suggestion
- Pacification
- Pressure
- Exhaustion
- Distraction

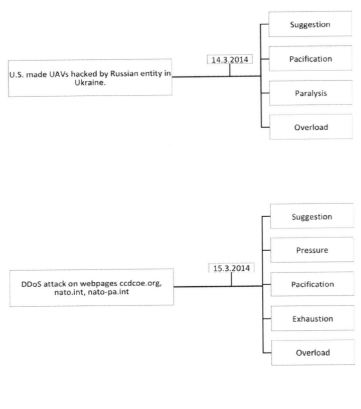

U.S. made UAVs hacked by Russian entity in Ukraine.

14.3.2014

- Suggestion
- Pacification
- Paralysis
- Overload

DDoS attack on webpages ccdcoe.org, nato.int, nato-pa.int

15.3.2014

- Suggestion
- Pressure
- Pacification
- Exhaustion
- Overload

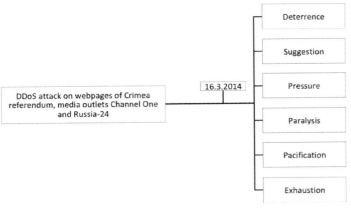

DDoS attack on webpages of Crimea referendum, media outlets Channel One and Russia-24

16.3.2014

- Deterrence
- Suggestion
- Pressure
- Paralysis
- Pacification
- Exhaustion

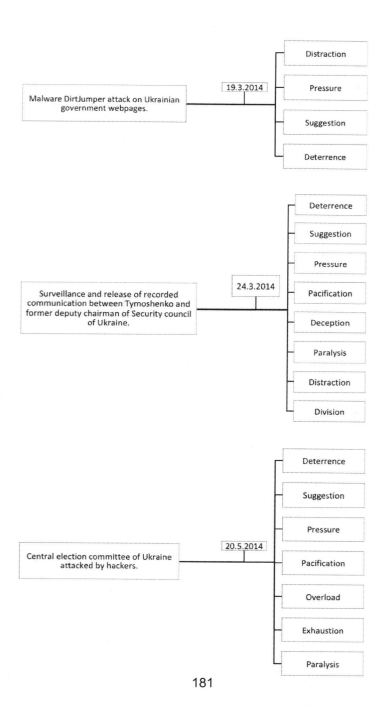

Malware DirtJumper attack on Ukrainian government webpages. — 19.3.2014
- Distraction
- Pressure
- Suggestion
- Deterrence

Surveillance and release of recorded communication between Tymoshenko and former deputy chairman of Security council of Ukraine. — 24.3.2014
- Deterrence
- Suggestion
- Pressure
- Pacification
- Deception
- Paralysis
- Distraction
- Division

Central election committee of Ukraine attacked by hackers. — 20.5.2014
- Deterrence
- Suggestion
- Pressure
- Pacification
- Overload
- Exhaustion
- Paralysis

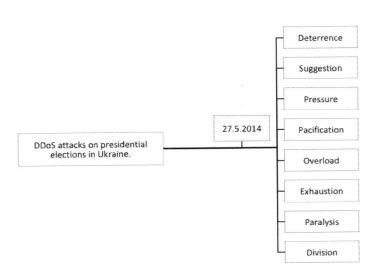

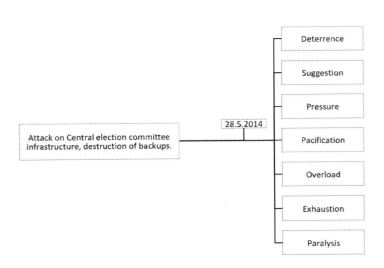

182

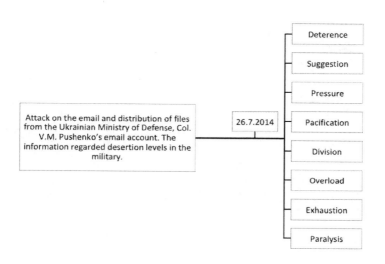

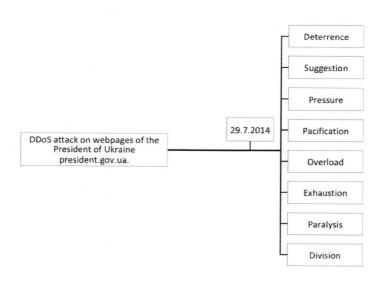

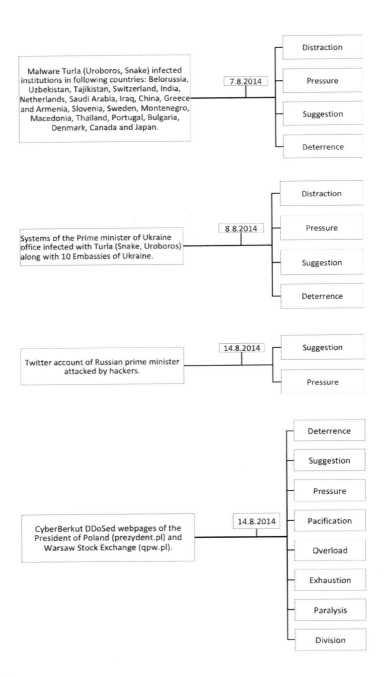

Malware Turla (Uroboros, Snake) infected institutions in following countries: Belorussia, Uzbekistan, Tajikistan, Switzerland, India, Netherlands, Saudi Arabia, Iraq, China, Greece and Armenia, Slovenia, Sweden, Montenegro, Macedonia, Thailand, Portugal, Bulgaria, Denmark, Canada and Japan.

7.8.2014

- Distraction
- Pressure
- Suggestion
- Deterrence

Systems of the Prime minister of Ukraine office infected with Turla (Snake, Uroboros) along with 10 Embassies of Ukraine.

8.8.2014

- Distraction
- Pressure
- Suggestion
- Deterrence

Twitter account of Russian prime minister attacked by hackers.

14.8.2014

- Suggestion
- Pressure

CyberBerkut DDoSed webpages of the President of Poland (prezydent.pl) and Warsaw Stock Exchange (qpw.pl).

14.8.2014

- Deterrence
- Suggestion
- Pressure
- Pacification
- Overload
- Exhaustion
- Paralysis
- Division

184

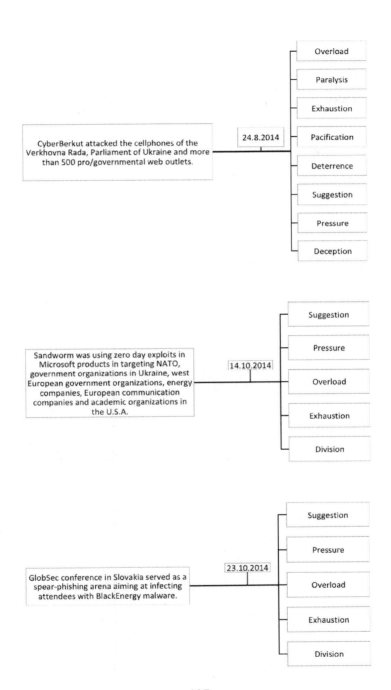

		Overload
CyberBerkut attacked the cellphones of the Verkhovna Rada, Parliament of Ukraine and more than 500 pro/governmental web outlets.	24.8.2014	Paralysis
		Exhaustion
		Pacification
		Deterrence
		Suggestion
		Pressure
		Deception

		Suggestion
Sandworm was using zero day exploits in Microsoft products in targeting NATO, government organizations in Ukraine, west European government organizations, energy companies, European communication companies and academic organizations in the U.S.A.	14.10.2014	Pressure
		Overload
		Exhaustion
		Division

		Suggestion
GlobSec conference in Slovakia served as a spear-phishing arena aiming at infecting attendees with BlackEnergy malware.	23.10.2014	Pressure
		Overload
		Exhaustion
		Division

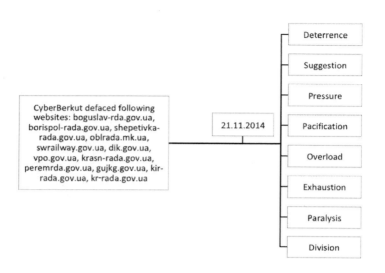

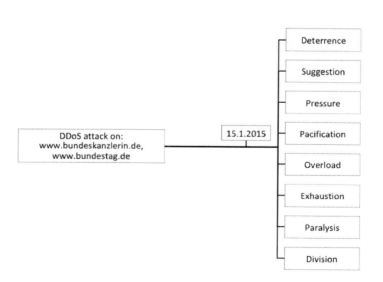

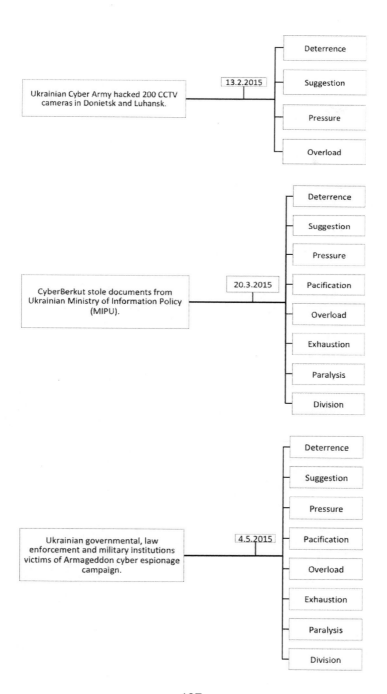

Ukrainian Cyber Army hacked 200 CCTV cameras in Donietsk and Luhansk.

13.2.2015

- Deterrence
- Suggestion
- Pressure
- Overload

CyberBerkut stole documents from Ukrainian Ministry of Information Policy (MIPU).

20.3.2015

- Deterrence
- Suggestion
- Pressure
- Pacification
- Overload
- Exhaustion
- Paralysis
- Division

Ukrainian governmental, law enforcement and military institutions victims of Armageddon cyber espionage campaign.

4.5.2015

- Deterrence
- Suggestion
- Pressure
- Pacification
- Overload
- Exhaustion
- Paralysis
- Division

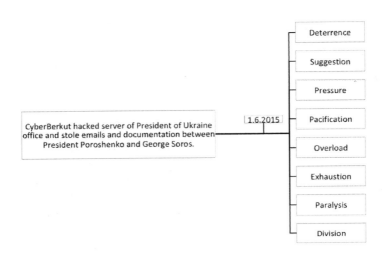

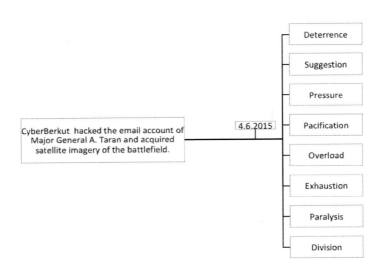

MILITARY DECISION-MAKING

According to the Russian military (and proven by the analysis of doctrinal documents in the previous chapters), superiority in the revolution of military affairs proceeds from an advantage in "information warfare" which Fitzgerald201 regarded in 1997 as reconnaissance, surveillance, and target acquisition systems, along with "intelligent" command and control systems. The "intelligence" of systems does not refer to autonomous or artificial intelligence, or machine learning concepts, but to the intellectualization of previously decentralized systems of command and control. The precision strike allowed by the digitalization (informationalization) of command and control was largely made possible due to the "intellectualization" of the command and control, reconnaissance and targeting systems. Intellectualization allowed for reduction of time in passing the commands and information between commanders and the troops. It also limited the number of necessary human interlocutors in passing the information. Command and control structures could therefore make almost real-time decisions and having computer-based calculations and coordination of forces.202 The effectiveness achieved by the shortened cycle of military decision-making, however, created a new area of potential vulnerabilities exploitable by information-technical and information-psychological assets.

The information warfare necessary for achieving military objectives has several components:203

a. Acquisition of information regarding the adversary and the troops in the field, information processing and incorporation within the command and control functions of the military operation.

b. Denying the opponent the acquisition, processing and exchange of information on his own troops and conditions of the operation, and actively utilizing means to disrupt the opponent's ability to gain the full picture of the situation, including distorting sensory awareness, be it technical or psychological classification.[204]

c. Defending the networks, sensors and command and control structure from the B component, conducted by the opponent against your decision-making cycle and information processing.

In the realm of the digitalized battlefield, component B can be viewed in today's terms as offensive cyber capabilities, with component C regarded as cybersecurity, cyber defense or information assurance protection.

The impact on the strike-reconnaissance complex is therefore not only on the technological aspect of the decision-making system but also at the highly vulnerable human element of the decision maker himself, be it the commander on the tactical, operational or strategic level.

As pointed out by Fitzgerald:

> The ultimate objective of information warfare (in military terms) is to achieve in-

formation dominance over the opponent, where the information control over one's own troops and weapon control organs is more complete, precise, reliable, and timely than the opponents' corresponding control organs (decision-making cycle). Thus, the Russians define information warfare as a complex of measures for information support, information opposition and information defense conducted according to a single concept and plan in order to seize and maintain information dominance over the opponent in the preparation and course of combat actions.[205]

Modern strike-reconnaissance complexes are reliant on technology and maintaining technical dominance is imperative in pursing military objectives. It also allows exploitation in situations when the conventional military might is not comparable, meaning one side is weaker in the number of troops, weapon platform efficiency, or restrained in the realm of target acquisition.

To scale back from the military operation itself, and the employment of information-technical and information-psychological assets on the command infrastructure, it is important to note that control over the civilian decision-making process, through disinformation and influence campaigns, or disrupting the communication between civilian leadership and the military leadership, can achieve the same results as if debilitating the enemy forces physically. The difference lies in not allowing the situation to escalate to hostilities. Military doctrines elaborate on the notion of compromising the decision-making mechanism on the civilian side of the leadership by efficiently denying the utilization of military assets in pursuing strategic objectives such as defending state sovereignty.

Elements necessary to protect for effective exercise of command and control are the sensory awareness assets such as radar and sensors for speed, altitude, humidity, elevation as well as electromagnetic spectrum detectors and measurement devices, and radiation sensors. This list also includes communication lines, relay stations, and satellites allowing communication and geographical location.

All these components of command and control are digitalized, vulnerable to information-technical attacks, or by the U.S./European-centric jargon – cyber and electronic warfare. Among one of the striking examples of an information-technical attack is 1990s First Gulf war debilitation of the Iraqi air defense systems during Operation Desert Storm.[206] The air defense systems were infected with logical bombs giving the United Nations-sanctioned forces air superiority which lead to taking out entire command and control structures within the Iraqi military and civilian leadership.

Superiority in reconnaissance, command and control, and electronic warfare is said to be the main factor in raising the qualitative indices of weapons and military equipment, which will have a "decisive" effect on the course and outcome of combat operations. Under all circumstances the side that has advantages in these areas will always possess greater capabilities, even if the other side has definite advantages in nuclear and, even more so, conventional weapons.[207]

OBSERVATIONS

Understanding the strategic mindset of the opponent is essential. Preconditions are derived from such a mindset that frames the perception of threats, which then defines the security and defense posture. To better understand the background, several red lines – or as stated below, "must-nots" – are presented.

The "must-nots" are put in the context of information warfare. These "must-nots" were presented [208] by a leading scholar and practitioner[209] on Russian strategic behavior and defense posture as red lines for the regime in Russia as not to be crossed to ensure the survival of the establishment.

"Endangered regime"

Any intervention in Russian domestic affairs perceived as potentially threatening the stability of the regime and its survival is a red line. This clearly also regards the information sphere and the national security and strategic posture presented in the documents analyzed in the previous chapters. Any information-related intervention is deemed as a threat. Example of such an informational risk is Elon Musk's Starlink project – with two demo satellites, called Tintin A and Tintin B already in orbit – that aims at providing global broadband service.210 This is contradictory to the efforts of the Russian establishment to have the information sphere, infrastructure and information itself under control.

"Neglecting of Russia and its interests by external players."

Russia, as stated in the doctrinal documents analyzed in this study, perceives that it is facing information warfare. This is done by undermining its domestic and international policies through coordinated efforts in the information sphere. As elaborated on by the undisclosed expert, Russia seeks to have the perception of its might and its role as a superpower preserved. In fact, the ability to project power and pursue its interests it is of strategic importance. The information campaign Russia is allegedly subjected to has the potential of weakening the regime domestically and internationally, leading to the neglect of Russia's interest through an undermined position in the international community.

"Deprivation of Russia of opportunity to act, restricting it of freedom of action."

From the perspective of information warfare, influence activities and deception, the restriction can be achieved by two measures: information-technical and information-psychological.

Information-technical measures when employed at the command and control and communication infrastructure level have the capacity to debilitate any activity. These measures also disrupt the integrity debilitating the capacity to act, on the technical level of decision-making such as the capacities and communications.

Information-psychological measures employed against the population and decision makers to influence or otherwise hamper the capacity of the regime to gain widespread support for its activities, or manipulation of the decision-making cycle to such an extent that the decision cannot be made in compliance with the strategic interests of the Russian Federation. This can be done, for instance, through reflexive control when Russian decision makers are faced with challenges and events that do not allow them freedom of action, but lead to a pre-set outcome, defined by external actors. Losing the initiative, maneuverability and having to react to predetermined conditions and consequences is a must-not scenario for Russian leadership.

"Losing the right to influence decisions and condition setting."

Another red line for the Russian establishment is losing the right to influence decisions, such as the veto power in the Security Council of the United Nations. Apart from the importance of this tool in shaping global affairs and international security, losing this right would lead to restricted opportunity and end in being subjected to a predisposed set of events by external actors. The connection with information warfare, influence campaigns, and deception in cyberspace is as follows: control over the perception of Russia among the population, decision-makers and the international community is of strategic importance. A weakened position of the Russian Federation through the information sphere, leading to an adjusted perception within the international community, is undesired. The perception of power and power projection are based on available information and the ability to construct reality. Therefore, it is subject to the control of information resources and content management. Manipulating sensory awareness, information resources and shaping the picture of Russia as a global actor and superpower have the power to alter the power balance. This is related to the notion of information superiority also presented in the study.

As noted in previous chapters, Russian understanding of cybersecurity differs from the U.S.-European infrastructure-centric perspective. One of the defining moments for the Russian national security posture and military thought was Operation Desert Storm.

"During operation Desert Storm in 1991, for the first time, electronic warfare and electronic countermeasures were the equivalent of conventional fire power in effectiveness. All the most important enemy targets were continuously subjected to electronic-fire pressure which disrupted the command-and-control and communications system simultaneously at all command echelons from tactical to strategic. Fourth, electronic and fire strikes were precisely coordinated by objective, place, and time. By being combined, they conjointly complemented and strengthened each other."[211]

That was 27 years ago and the core of the measures was in information-technical measures.

Over the past decades information-psychological measures have been gaining prominence and according to Roy Godson:[212]

"In the final years of the Soviet Union there was enough information on their active measures systems to conclude that approximately 15,000 personnel and several billions of hard currency annually were being spent on these activities — aimed mostly at the U.S. and its allies."

One should ask about the utilization of all the intellectual capacity, apparatus and experience today. The victims of today, subject to psychological and technical information measures of deception and influence, lack the intellectual and institutional framework to address operations conducted in the information sphere, in cyberspace. Recommendations have been made over the past few years regarding how to respond and act. However, those presented by Roy Godson,[213] should be analyzed and properly implemented. In the testimony in front of the Senate Select Committee on Intelligence on March 30th, 2017 he provided the audience with *"what is to be done,"*[214] perhaps teasingly referring to the title of pamphlet[215] of Vladimir Lenin elaborating on the prominent position of deception in Soviet thinking. The recommendations are related to the Russian Federation, but are in fact universal in countering deception and information operations.

What is ostensibly present is the lack of ambition, lack of leadership vision and ability to drive the change in responding to the challenges emanating from the exploitation of cyberspace. Among the recommendations made is the need for developing a strategic approach. The following measures are subsets, applicable to strategic, operational and tactical level/individual levels:

- Create an intellectual basis sufficient for understanding the strategic intent, objectives and information technical and information psychological means altogether.
- Define red lines not to be crossed by an adversary.
- Adopt deterrence mechanisms and tools to prevent the crossing of the red lines.
- Convey a message of resolve and readiness to apply above mentioned.
- Seize the initiative and refrain from reactive postures.
- Pre-plan responses and contingency plans to information-technical and information-psychological measures employed against the strategic interests, and having an impact on, national security and defense capabilities.
- Actively and assertively oppose activities and operations contradicting the values and narratives natural to the society and its allies.
- Educate decision makers, individuals and the general population in the realm of critical thinking and cognitive and psychological prerequisites for the analysis

and perception of reality. If the targeted individuals are not familiar with the principles of deception and resilience, they can hardly notice they are being deceived or subjected to an information technology related attack.

- Create an institutional framework and capacity to analyze, understand and devise countermeasures on all levels of governance. The interagency "Active Measures Working Group," based first in the State Department and later in the U.S. Information Agency can be a historical example to draw lessons from.[216]

- Reduce the effectiveness of information operations and campaigns by exposing the strategic intent, plans, and activities of the potential adversary.

- Disseminate the positive narratives contradicting the narratives disseminated by the potential opponent.

- Sensitize policy and decision makers, commanders, analysts and the general population to the potential of cyberspace and related tools and activities in influencing perception and altering decision-making algorithms. The ability to interpret and process information, and analyze and adapt one's behavior is vital for making decisions.

- Ensure that training and exercise for the military commanders, troops, but also decision makers lead to: *"Strong nervous system, quick thinking, quick orientation to the environment, quick logical conclusion capacity, sense of responsibility, coolness under pressure, ability to shift attention quickly."*[217]

One of the critical issues to address is the lack of understanding among the population, including leadership, of the potential uses of modern technologies and how quickly deceived one can be through everyday consumption of convenient services and technologies. Making people aware leads to higher resilience when potential risks are understood, and people are thus less susceptible to information operations. How is the individual supposed to be resilient and secure himself, if he has no understanding of what he is facing?

The critical component of any recommendation is, however, the individual. Individuals are the common denominator of the processes and decision-making, and are the ultimate targets of information-psychological and, in case the aim is not the infrastructure itself, information-technical attacks. If the individual is resilient, then the activities he participates in are resilient, be it analysis, information processing or decision-making. The individual's capacity to understand the complexities of the modern world shields him or her from sensory awareness tampering. This understanding and the ability to critically assess helps him be resilient towards cognitive biases.

It is far easier to lead a target astray
by reinforcing the target's existing beliefs,
thus causing the target to ignore the contrary
evidence of one's true intent,
than it is to persuade a target to change
his or her mind. [218]

From the perspective of NATO, presented by Can Kasapoglu,[219] "NATO needs new intelligence analysis and strategic forecasting capabilities for getting a grip on the new Russian challenge." However, the Russian challenge of employing reflexive control is not new, he continues: "First and foremost, the North Atlantic Alliance strategic community should recognize the Russian 'reflexive control' campaign that could bring about a menacing 'analytical paralysis' when assessing Moscow's true intentions..."[220]

The analytical paralysis, as he states, "may well include 'buying' the less risky options that the Russian elite offer in order to pursue a policy which might be seen 'carefully-calculated and risk-aversive,' but unintentionally paving the ground for more gains for Moscow through use or threat of force."[221] Kasapoglu concludes that it is important for "the North Atlantic Alliance to promote necessary institutions and concepts to develop a thorough understanding of hybrid warfare, and the Russian interpretation of it, in the form of reflexive control-driven nonlinear warfare."[222]

The key terms that should be underlining any efforts are preemptive understanding, and strategic initiative and planning, as opposed to short term reactive activity. It is important to note that the perception of highly orchestrated and organized strategy, involving strategic, operational and tactical layers of deception, reflexive control and information operations across the Russian government and coordinated with proxy actors provides a perception of a mighty force to be reckoned with. However, this could be part of the deception campaign itself, to project a high degree of coordination and command and control of the adversarial activities. That does not mean that the activities should be underestimated, or may have less impact. On the contrary, if they are not entirely under control of a central body, the more dangerous they are with the lack of sufficient supervision and command and control.

CONCLUSION

Fitzgerald in 1997: "The most important objective of military conflicts in the near-term future may become affecting the psychology of the opponent – individual, collective, and mass."[223]

This book described the concepts of information superiority, reflexive control, and employment of past concepts of deception in the digitalized era. The amplification of information warfare methods in cyberspace, or from Russian perspective, the information sphere, was ushered in by the massive penetration of communication and information technologies underpinning the very existence of modern societies. Nevertheless, the evolution of technologies and their utilization call for the adoption of new techniques of influence, new command and control methods and military platforms. This will lead to an unforeseen level of exploitation of information-based societies and related vulnerabilities.

Influence operations, deception, and employment of information technical and information psychological measures allow the actor to avoid direct confrontation with its opponents. Also, operations in the information sphere enable the conduct of hostile activities under the threshold of armed conflict and the ability to probe the red lines of the opponent, all the while using plausible deniability to backed-escalate an initiative if necessary without repercussions in the physical domain. Operations in the information sphere also impact the resolve to resist.

The challenge is, however, the lack of understanding of the impact today. This study aimed to provide contextualization of applied concepts from the past, repurposed and adjusted to the digital present. Information weapons are serving a larger purpose. They not only inflict damage to the technical backbone of the information and communication infrastructure, but also have the potential to alter the balance of power and achieve strategic objectives.

The impact of information weapons can be immediate or cause effects over a more extended period, either by degrading the opponents' societal structures or striking at the moment when a pivotal decision has to be made. Information strikes can inflict damage and affect key individuals, the general population, and vulnerable societal groups. This can lead to the debilitation of societal functions, the collapse of institutions and eventually collapse of the state.

The critical component in winning a conflict where information weapons are employed is a thorough knowledge of the adversary gained by understanding his strategic objective, culture, operational code, capabilities – not only the military capabilities, but the culture, psyche, decision-making processes, and institutional frameworks. In essence, it is important to grasp how policymaking and stakeholders in society interact. Another no less critical component is the initiative. To be able not to behave according to predefined sets of analyzed behavior, to be able to exact some degree of surprise on the adversary, it is vital to maintain the initiative – to be the definer of events, not being reactive to them. If the entity allows itself only to react in this passive role, whether by lack of vision, leadership or objectives, it will enable the opponent to prepare the situations that require reaction according to anothers' interests and goals. As Fitzgerald states:[224] "A most important condition for the successful execution of psychological operations is considered to be constantly maintaining the offensive and holding the "psychological initiative."

LIST OF LITERATURE

Above Top Secret, (2014). *Read Why the USS Donald Cook Crapped It's Pants and was 'gravely demoralized' in the Black Sea*. [Online]. Available at: http://www.abovetopsecret.com/forum/thread104210 9/pg1 [Accessed 18th March 2018].

Beaumont, R. (1982). *Maskirovka: Soviet Camouflage, Concealment and Deception*. College Station, Texas: Center for Strategic Technology, A & M University System.

Bruusgaard, Ven K. (2016). *Global Politics and Strategy August–September 2016*. IISS. Survival. Global Politics and Strategy, [Online]. Available at: https://www.iiss.org/en/publications/survival/sections/2016-5e13/survival--global-politics-and-strategy-august-september-2016-2d3c/58-4-02-ven-bruusgaard-45ec [Accessed: 7th June 2016].

Business Insider, (2014). *Putin Has Taken Control Of Russian Facebook*. [Online]. Available at: http://www.businessinsider.com/putin-has-taken-control-of-russian-facebook-2014-4 [Accessed 18th March 2018].

Caddell, J. W. (2004). *Deception 101-Primer on deception*. Carlisle Barracks, PA: Strategic Studies Institute, U.S. Army War College.

Calhoun, L. (2018). Elon Musk Just Sent More Stuff Into Space--This Time, It's Even Better Than the Roadster. *Inc.* [Online]. Available at: https://www.intelligence.senate.gov/sites/default/files/documents/os-rgodson-033017.pdf [Accessed: 28th January 2018]

Carr, J. and Dao, ´D. (2011). „*4 Problems with China and Russia's International Code of Conduct for Information Security, and Dao, ´D. and Giles, K. (2011). 'Russia's Public Stance on Cyberspace Issues* ´, *in* Czosseck,Ch. Ottis, R. and Ziolkowski, K.

(2012). 4th International Conference on Cyber Conflict. Tallin : NATO CCD COE Publications.

Corman, S. R., and Dooley, K. J. (2009). *Strategic Communication on a Rugged Landscape. Principles for Finding the Right Message.* Consortium for Strategic Communication, Arizona State University. Tucson.

CyberBerkut (2014). *CyberBerkut suspended the operation of the Ukrainian Central Election Commission* [Online]. Available at: https://cyber-berkut.org/en/olden/index2.php [Accessed 20th January 2017].

CyberBerkut (2014). *E-mail of the Ukrainian Ministry of Defense colonel has been hacked* [Online]. Available at: https://cyber-berkut.org/en/olden/index2.php [Accessed 20th January 2017].

CyberBerkut (2014). *New punishers' losses data in the South-East.* [Online]. Available at: https://cyber-berkut.org/en/olden/index2.php [Accessed 20th January 2017].

CyberBerkut (2014). *CyberBerkut hacked Kiev digital billboards.* [Online]. Available at: https://cyber-berkut.org/en/olden/index2.php [Accessed 20th January 2017]

Dailey, B. D. and Parker P. (1987). *Soviet Strategic Deception.* Stanford: Hoover Institution Press. ISBN 066913208X.

Daniel, D. C., and Herbig, L. K. (2013). *Strategic Military Deception: Pergamon Policy Studies on Security Affairs.* New York: Elsevier. p. 392. ISBN 1483190064.

DEF CON (also written as DEFCON, Defcon, or DC), one of the world's largest hacker conventions. [Online]. Available at: https://www.defcon.org/ [Accessed 20th January 2017].

Digital Forensic Research Lab, (2017). *Russia's Fake "Electronic Bomb". How a fake based on a parody spread to the Western mainstream.* [online]. Available at: https://medium.com/dfrlab/russias-fake-electronic-bomb-4ce9dbbc57f8 [Accessed 18th March 2018].

DW News, (2018). *Russia moves toward creation of an independent internet.* [Online]. Available at: http://www.dw.com/en/russia-moves-toward-creation-of-an-independent-internet/a-42172902 [Accessed 18th March 2018].

EFJ, (2017). *Russian 'foreign agents' media law threatens media freedom. [Online].* Available at: https://europeanjournalists.org/blog/2017/11/28/russian-foreign-agents-media-law-threatens-media-freedom/ [Accessed 18th March 2018].

Erger, A. (2005). Yoda and the Jedis: The Revolution in Military Affairs and the Transformation of War. *The OST's Publication on Science & Technology Policy.* [Online] 7. Available at: http://www.ostina.org/content/view/274/ [Accessed 20th January 2017].

Fink, A. L. (2017). *The Evolving Russian Concept of Strategic Deterrence: Risks and Responses.* Arms Control Association. [Online]. Available at: https://www.armscontrol.org/act/2017-07/features/evolving-russian-concept-strategic-deterrence-risks-responses [Accessed: 28th January 2018].

Fitzgerald, M. C. (1999). *Russian Views on IW, EW, and Command and Control: Implications for the 21st Century.* [Online]. Available at: http://www.dodccrp.org/events/1999_CCRTS/pdf_files/track_5/089fitzg.pdf [Accessed: 28th January 2018].

FOX NEWS, (2017). *Russia claims it can wipe out US Navy with single 'electron-*

ic bomb'. [Online]. Available at: https://web.archive.or
g/web/20170517 23954/http://www.foxnews.com/wor
ld/2017/04/19/russia-claims-it-can-wipe-out-us-navy-
with-single-electronic-bomb.html [Accessed 18th
March 2018].

Gilbert, D. T. (1991). *How Mental Systems Believe*.
American Psychologist. [Online] 46. Available at: http://www.danielgilbert.com/Gillbert%20(Ho
w%20Mental%20Systems%20Believe).PDF [Accessed:
28th January 2018].

Giles, K. (2011). *"Information Troops" - A Russian
Cyber Command?*, in Cyber Conflict (ICCC), 2011 3rd
International Conference on, Tallin: IEEE, pp. 43-59.
ISBN 9781612842455.

Godson, R. (2017). *Written Testimony of ROY
GODSON to the Senate Select Committee on Intelligence, Open Hearing, March 30, 2017 "Disinformation: A Primer in Russian Ac-
tive Measures and Influence Campaigns."* [Online]. Av
ailable at: https://www.intelligence.senate.gov/sites/d
efault/files/documents/os-rgodson-033017.pdf
[Accessed: 28th January 2018].

Green, W. C. and W. R. R. (1993). Marshals of the
Soviet Union A. A. Grechko and N. V. Ogarkov
[successive Chairmen of the Main Editorial
Commission], *The Soviet Military Encyclopedia*,
English Language Edition, Vol. 1, pp. 345-346,
Westview Press, Boulder.

Heickerö, R. (2010). *Emerging cyber threats and
Russian views on Information warfare and Infor-
mation operations*. Defence Analysis, Swedish Defence
Research Agency (FOI). [Online]. Availa-
ble at: http://www.highseclabs.com/data/foir2970.pdf
[Accessed: 28th January 2018].

Hogan, H. (1967). *Lenin's Theory of Reflection*.
Master's Thesis. McMaster University.

Human Rights First, (2017). *Russian Influence in Europe. [Online].* Available at: https://www.humanrightsfirst.org/resource/russian-influence-europe [Accessed 18th March 2018].

lnr.todays, (2015). *В аэропорту Луганска найдены "Ст*

ингеры" (оперативная съемка) #ЛНРсегодня

. [online video] Available athttps://www.youtube.com/watch?v=mJr7zUXwBx8&feature=youtu.be [Accessed 18th March 2018].

Inform Napalm, (2016). *Russian Leer-3 EW system revealed in Donbas.* [online]. A vailable at: https://informnapalm.org/en/russian-leer-3wf-donbas/ [Accessed 18th March 2018].

In-fo Wars, (2014). Russians Disable U.S. Guided Missile Destroy-er. [Online]. Available at: https://www.infowars.com/russians-disable-u-s-guided-missile-destroyer/ [Accessed 18th March 2018].

Ionov, M.D. (1995). On Reflective Enemy Control in a Military Conflict. Military Thought. English edition. p.45-50.

Kasapoglu, C. (2015). Russia ́s Renewed Military Thinking: Non-Linear Warfare and Reflective Control. *Research Paper. Rome: Research Division – NATO Defence College.* [Online] 121. Available at: http://cco.ndu.edu/Portals/96/Documents/Articles/russia%27s%20renewed%20Military%20Thinking.pdf [Accessed: 28th October 2017].

Komov, S. A. (1997). About Methods and Forms of Conducting Information Warfare. *Military Thought*, 4, 18-22.

Korotchenko, Y. G. (1996). Information-Psychological Warfare in Modern Conditions. Military Thought. English edition. p. 22-27.

Kragh, M. and Åsberg, S. (2017). Russia's strategy for influence through public diplomacy and active measures: the Swedish case. *Journal of Strategic Studies*, 40.6, 773-816.

Kramer, D. F., Starr, S. and Wentz, H. L. (2009). *Cyberpower and National Security.* Dulles : Potomac Books, Inc. p. 664. ISBN 1597979333.

Lenin, V. I. (1902). *What Is To Be Done?.* Marxists Internet Archive [Online]. Available at: https://www.marxists.org/archive/lenin/works/download/whatitd.pdf [Accessed: 28th January 2018].

Lexpress, (2015). *Piratage de TV5 Monde: l'enquête s'oriente vers la piste russe.* [Online]. Available at: https://www.lexpress.fr/actualite/medias/piratage-de-tv5-monde-la-piste-russe_1687673.html [Accessed 18 March 2018].

Logvinov, A. (2015). *Ukrainian rebels make fake video using weapons from the game 'Battlefield 3'.* Available at: https://meduza.io/en/lion/2015/07/23/ukrainian-rebels-make-fake-video-using-weapons-from-the-game-battlefield-3 [Accessed: 28th January 2018].

Mazarr, M. J. (2015). *Mastering the Gray Zone: Understanding a Changing Era of Conflict.* Carlisle: U.S. Army War College Carlisle.MCFaul, M. (2017) 6th January. Available at https://twitter.com/McFaul [Accessed: 18th March 2017].

Medvedev, S. A. (2015). *Offense-defense theory analysis of Russian cyber capability.* PhD Thesis. Naval Postgraduate School.

NATO Cooperative Cyber Defence Centre of Excellence, (2015). *An Updated Draft of the Code of Conduct Distributed in the United Nations – What's New?* [Online]. Available at: https://ccdcoe.org/updated-draft-code-conduct-distributed-united-nations-whats-new.html#footnote3_7g6pxbk [Accessed 18th March 2018].

O'Brien, T. N. (1989). *Russian Roulette: Disinformation in the U.S. Government and News Media. Master's Thesis.* South Carolina University Columbia.

Panagiotis, O. (2016). *Strategic Military Deception. Prerequisites of Success in Technological Environment.* Available at: http://www.rieas.gr/im ages/publications/rieas171.pdf [Accessed: 28 January 2018].

Radio Free Europe Radio Liberty, (2015). *Russian TV Deserters Divulge Details On Kremlin's Ukraine 'Propaganda'* [Online]. Available at: https://www.rferl.org/a/rus sian-television-whistleblowers-kremlin-propaganda/27178109.html [Accessed 18th March 2018].

Reid, C. (1987) "Reflexive Control in Soviet Military Planning,". In Dailey, B. D. and Parker P. *Soviet Strategic Deception.* Stanford: Hoover Institution Press.

Reuters, (2017). Russia's RT America registers as 'foreign agent' in U.S. [Online]. Available at: https://www.reuters.com/article/us-russia-usa-media-restrictions-rt/russias-rt-america-registers-as-foreign-agent-in-u-s-idUSKBN1DD25B [Accessed 18th March 2018].

RG RU, (2016). Что напугало американский эсмин ец. [Online]. Available at: https://rg.ru/2014/04/30/reb-site.html [Accessed 18th March 2018].

Rothstein, H., and Whaley, B. (2013). The Art and Science of Military Deception (Artech House Intelligence and Information Operations). New York: Artech House.

Russian Federation. President of the Russian Federation (2000). *Information Security Doctrine of the Russian*

Federation.[Online]. Available at:
https://info.publicintelligence.net/RU-
InformationSecurity-2000.pdf [Accessed 20th January
2017].

Russian Federation. President of the Russian Federation (2010). *Military doctrine of the Russian Federation.* [Online]. Available at:http://carnegie endowment.org/files/2010russia_military_doctrine.pdf [Accessed 20th January 2017].

Russian Federation. President of the Russian Federation (2010). *Russian Federation Armed Forces' Information Space Activities Concept.* [Online]. Available at:
http://eng.mil.ru/en/science/publications/more.htm?i d=10845074@cmsArticle [Accessed 20th January 2017].

Russian Federation. President of the Russian Federation (2014). *The Military Doctrine of the Russian Federation.* [Online]. Available at:
https://rusemb.org.uk/press/2029 [Accessed 20th January 2017].

Russian Federation. President of the Russian Federation(2015). *Doctrine of information Security of the Russian Federation*

Russian Federation. President of the Russian Federation(2015). *Russian Federation National Security Strategy.* [Online]. Available at: http://www.ieee.es/Galerias/fichero/OtrasPubli caciones/Internacional/2016/Russian-National-Security-Strategy-31Dec2015.pdf [Accessed 20th January 2017].

Russian Federation. President of the Russian Federation (2016). *Doctrine of Information Security of the Russian Federation* [Online]. Available at:
http://www.mid.ru/en/foreign_policy/official_docum

ents/ /asset_publisher/CptICkB6BZ29/content/id/25
63163 [Accessed 20th January 2017].

Russian Military Review, (2015). *День инноваций
ЮВО: комплекс РЭБ РБ-341В «Леер-3»*. [online]. A
vailable at: https://archive.is/MDecB [Accessed 18th
March 2018].

RT, (2015). *NATO trace 'found' behind witch hunt
website in Ukraine*. [online].
Available at: https://www.rt.com/news/253117-nato-
ukraine-terrorsite/ [Accessed 18th March 2018].

Schoen, F. and Lamb, C. (2012). *Deception, disinformation, and strategic communications*. Washington D.C.: National Defense University Press.

Soldatov, A. (2014). Russia's communications interception practices (SORM) [presenta-
tion] *Agentura.Ru*. [Online]. Available at: http://www.
eur-
parl.europa.eu/meetdocs/2009_2014/documents/libe
/dv/soldatov_presentation_/soldatov_presentation_e
n.pdf [Accessed: 7th June 2017].

Sputnik News (2014). *Russische SU-24 legt amerikanischen Zerstörer lahm*. [Online]. Available at:
https://de.sputniknews.com/meinungen/2014042126
8324381-Russische-SU-24-legt-amerikanischen-
Zerstrer-lahm/ [Accessed 18th March 2018].

Taham, S. (2013). *U.S. Governmental Information
Operations and Strategic Communications: A
Discredited Tool Or User Failure? : Implications for
Future Conflict*. Carlisle Barracks, PA: Strategic
Studies Institute, U.S. Army War College. p. 80. ISBN
158487600X.

Tottenkoph, R. (2016). Hijacking the Outdoor
Digital Billboard [PowerPoint presentation] *DEF CON
Hacking Conference*. [Online]. Available at:
https://www.defcon.org/images/defcon-16/dc16-

presentations/defcon-16-tottenkoph-rev-philosopher.pdf [Accessed: 7th June 2016].

The Moscow Times, (2014). *Vkontakte Founder Says Sold Shares Due to FSB Pressure*. [Online]. Available at: https://themoscowtimes.com/news/vkontakte-founder-says-sold-shares-due-to-fsb-pressure-34132 [Accessed 18th March 2018].

The New York Times, (2015). *The Agency*. [Online]. Available athttps://www.nytimes.com/2015/06/07/magazine/the-agency.html [Accessed 18th March 2018].

The list of backbone organizations, approved by the Government Commission to improve the sustainability of the development of the Russian economy [Online]. Available at: https://web.archive.org/web/20081227071316/http:/www.government.ru/content/governmentactivity/mainnews/33281de212bf49fdbf39d611cadbae95.doc [Accessed 18th March 2018].

Thomas, L. T. (1996). Russian Views on Information-Based Warfare. *Airpower Journal*, Special. Edition, 25-35.

Thomas, Timothy L. "The Russian View Of Information War." *The Russian Armed Forces at the Dawn of the Millennium* (2000): 335.

Thomas, L. T. (2004). Russia's reflexive control theory and the military. *Journal of Slavic Military Studies*, 17.2, 237-256.

Thomas, Timothy L. *Russian information warfare theory: The consequences of August 2008*. na, 2010.

Thomas, T. L. (2010). *Russian information warfare theory: The consequences of August 2008*. in *The Russian Military Today and Tomorrow: Essays in Memory of Mary Fitzgerald*, ed. Blank, J. S. and Weitz, R. Carlisle, PA: Strategic Studies Institute, pp. 265-301, ISBN 158487449X.

Trend Micro, (2016). *Operation Pawn Storm: Fast Facts and the Latest Developments.* [Online]. Available at: https://www.trendmicro.com/vinfo/us/security/news/cyber-attacks/operation-pawn-storm-fast-facts [Accessed 18 March 2018].

United States. Department of State (1989). *Soviet influence activities : a report on active measures and propaganda, 1987-1988.* U.S. Dept of State, Washington.

U. S. Department of Defense, (2014). *Russian Aircraft Flies Near U.S. Navy Ship in Black Sea.* [Online]. Available at: http://archive.defense.gov/news/newsarticle.aspx?id=122052 [Accessed 18th March 2018].

Vesti News, (2017). Electronic Warfare: How to Neutralize the Enemy Without a Single-Shot. [online video] Available at: https://www.youtube.com/watch?v=vI4uS307ydk&feature=youtu.be [Accessed 18 March 2018].

VOA NEWS, (2017). *Sinister Text Messages Reveal High tech Front in Ukraine War.* [online]. A vailable at: https://www.voanews.com/a/sinister-text-messages-high tech-frony-ukraine-war/3848034.html [Accessed 18th March 2018].

VOA NEWS, (2018). *Russia's Foreign Agent Law Has Chilling Effect On Civil Society Groups, NGOs.* [Online]. Available at: https://www.voanews.com/a/russia-labels-media-outlets-as-foreign-agents/4221609.html [Accessed 18th March 2018].

Voltaire Network, (2016). *About Voltaire Network.* [Online]. Available at: http://www.voltairenet.org/article150341.html [Accessed 18th March 2018].

Washington Post, (2014). Putin says Russia will protect the rights of Russian abroad. [Online]. Available at: https://www.washingto

npost.com/world/transcript-putin-says-russia-will-protect-the-rights-of-russians-abroad/2014/03/18/432a1e60-ae99-11e3-a49e-76adc9210f19_story.html?utm_term=.6d532d8be341 [Accessed 18th March 2018].

Washington Post, (2017). Fingered for Trading in Russian Fake News. [Online]. Available at: https://www.washingtonpost.com/blogs/erik wemple/wp/2017/06/07/foxnews-com-fingered-for-trading-in-russian-fake news/?utm_term=.647f82ed21dc [Accessed 18th March 2018].

Weiss, G. W. (1996). The Farewell Dossier. Dumping the Soviets. *Studies in Intelligence. Central Intelligency Agency.* [Online] 39, 5. Available at: https://www.cia.gov/library/center-for-the-study-of-intelligence/csi-publications/csi-studies/studies/96unclass/farewell.htm [Accessed 20th January 2017].

Украина Сегодня

, (2015). *'Cyberberkut' hacked Kyiv billboards* . [Online]. Available at: https://www.youtube.com/watch?v=E8A2MIkiavE [Accessed 18th March 2018].

NOTES

[1] Mazarr, M. J. (2015). *Mastering the Gray Zone: Understanding a Changing Era of Conflict*. Carlisle: U.S. Army War College Carlisle.

[2] Caddell, J. W. (2004). *Deception 101-Primer on deception*. Carlisle Barracks, PA: Strategic Studies Institute, U.S. Army War College.

[3] Caddell, J. W. (2004). *Deception 101-Primer on deception*. Carlisle Barracks, PA: Strategic Studies Institute, U.S. Army War College.

[4] Rothstein, H., and Whaley, B. (2013). *The Art and Science of Military Deception (Artech House Intelligence and Information Operations)*. New York: Artech House. OR Bacon, F. (1625). *Of Simulation and Dissimulation*. In Hawkins, M. J. (1973). Essays, London: J. M. Dent

[5] Rothstein, H., and Whaley, B. (2013). *The Art and Science of Military Deception (Artech House Intelligence and Information Operations)*. New York: Artech House.

[6] Caddell, J. W. (2004). *Deception 101-Primer on deception*. Carlisle Barracks, PA: Strategic Studies Institute, U.S. Army War College.

[7] Daniel, D. C., and Herbig, L. K. (2013). *Strategic Military Deception: Pergamon Policy Studies on Security Affairs*. New York: Elsevier.

[8] Rothstein, H., and Whaley, B. (2013). *The Art and Science of Military Deception (Artech House Intelligence and Information Operations)*. New York: Artech House.

[9] Caddell, J. W. (2004). *Deception 101-Primer on deception*. Carlisle Barracks, PA: Strategic Studies Institute, U.S. Army War College.

[10] Rothstein, H., and Whaley, B. (2013). *The Art and Science of Military Deception (Artech House Intelligence and Information Operations)*. New York: Artech House.

[11] Caddell, J. W. (2004). *Deception 101-Primer on deception*. Carlisle Barracks, PA: Strategic Studies Institute, U.S. Army War College.

[12] Caddell, J. W. (2004). *Deception 101-Primer on deception*. Carlisle Barracks, PA: Strategic Studies Institute, U.S. Army War College.

[13] Caddell, J. W. (2004). *Deception 101-Primer on deception*. Carlisle Barracks, PA: Strategic Studies Institute, U.S. Army War College.

[14] Caddell, J. W. (2004). *Deception 101-Primer on deception*. Carlisle Barracks, PA: Strategic Studies Institute, U.S. Army War College.

[15] Union A. A. Grechko and N. V. Ogarkov (1993). Marshals of the Soviet [successive Chairmen of the Main Editorial Commission], The Soviet Military Encyclopedia; English Language Edition, Vol. 1, William C. Green and W. Robert Reeves, ed. and trans., Boulder, CO: Westview Press, pp. 345-346.

[16] Medvedev, S. A. (2015). *Offense-defense theory analysis of Russian cyber capability*. PhD Thesis. Naval Postgraduate School.

[17] Thomas, T. L. (2010). *Russian information warfare theory: The consequences of August 2008.* in *The Russian Military Today and Tomorrow: Essays in Memory of Mary Fitzgerald*, ed. Blank, J. S. and Weitz, R. Carlisle, PA: Strategic Studies Institute, pp. 265-301.

[18] Thomas, T. L. (2010). *Russian information warfare theory: The consequences of August 2008.* in *The Russian Military Today and Tomorrow: Essays in Memory of Mary Fitzgerald*, ed. Blank, J. S. and

Weitz, R. Carlisle, PA: Strategic Studies Institute, pp. 265-301.

[19] Giles, K. (2011). *"Information Troops" - A Russian Cyber Command?*. Tallin: IEEE, pp. 43-59.

[20] Medvedev, S. A. (2015). *Offense-defense theory analysis of Russian cyber capability*. PhD Thesis. Naval Postgraduate School.

[21] Medvedev, S. A. (2015). *Offense-defense theory analysis of Russian cyber capability*. PhD Thesis. Naval Postgraduate School..

[22] Fitzgerald, M. C. (1999). *Russian Views on IW, EW, and Command and Control: Implications for the 21st Century*. [Online]. Available at: http://www.dodccrp.org/events/1999_CCRTS/pdf_files/track_5/089fitzg.pdf, [Accessed: 28th January 2018].

[23] *Ludendorff: Strategist* [online]. In:1992. Available http://www.dtic.mil/dtic/tr/fulltext/u2/a250915.pdf [Accessed 20th January 2017]

[24] Fitzgerald, M. C. (1999). *Russian Views on IW, EW, and Command and Control: Implications for the 21st Century*. [Online]. Available at: http://www.dodccrp.org/events/1999_CCRTS/pdf_files/track_5/089fitzg.pdf, [Accessed: 28th January 2018].

[25] Khalilzad, Zalmay, John P White, and Andy W. Marshall, eds (1999)., *Strategic Appraisal: The Changing Role of Information in Warfare*. [Online]. Available at https://www.rand.org/pubs/monograph_reports/MR1016.html. [Accessed: 28th January 2018].

[26] United States. Department of State (1989). *Soviet influence activities: a report on active measures and propaganda, 1987-1988*. U.S. Dept of State, Washington.

[27] In present day, it can be blogs, vlogs, online media outlets, forged digital documents, audio and video files, social media accounts and avatars. The role of media and delivery platforms of disinformation is reinforced today by online content and channels. However, the old-school journalists were targets long before cyber domain made it easier to create content and propagate it, or eventually target the media outlets. According to the testimony of Stanislav Levchenko, a KGB agent directly involved in influence campaigns on behalf of the Soviet Union: *"The Soviet leadership excellently understand that journalists are people who create public opinion in the West and are able to be used as agents of influence, - not directly, but in some kind of indirect way, because there are two types of agents of influence. The first is when a journalist or a businessman or some kind of political figure is recruited. The second is when he is used unconsciously, when they deliver to him a material or information favorable to the Soviet Union. I am speaking completely responsibly, because I, myself delivered such information to American, French, or German journalists and they used it successfully for the benefit of Soviet propaganda. It was not a lie. It was well prepared disinformation."* United States. Department of State (1989). *Soviet influence activities: a report on active measures and propaganda, 1987-1988.* U.S. Dept of State, Washington.

[28] In present day also focusing on domestic policies and domestic interests such as national security, democratic institutions or civic society.

[29] In present day also high level fora attracting witting and unwitting officials and dignitaries to project international support of the objectives of the fronts.

[30] Today Russian federation, and using fronts also covertly, in promoting the policies in favor of Russian interests, such as divisive political forces in European countries.

[31] Heickerö, R. (2010). *Emerging cyber threats and Russian views on Information warfare and Information operations.* Defence Analysis, Swedish Defence Research Agency (FOI). [Online]. Available at: http://www.highseclabs.com/data/foir2970.pdf [Accessed: 28th January 2018].

[33] Thomas, L. T. (2004). Russia's reflexive control theory and the military. *Journal of Slavic Military Studies,* 17.2, pp. 237-256.

[34] Ward, Amanda. *The Okhrana and the Cheka: Continuity and Change* [online]. 2014 [cit. 2018-01-30]. Available atat: : https://etd.ohiolink.edu/!etd.send_file?accession=ohi ou1398772391. Ohio University. [Accessed 20th January 2017].

[35] Giles, K. (2011). *"Information Troops" - A Russian Cyber Command?* Tallin: IEEE, pp. 43-59.

[36] Giles, K. (2011). *"Information Troops" - A Russian Cyber Command?* Tallin: IEEE, pp. 43-59.

[37] A comparison of the US and Societ economies: Evaluating the performance of the Soviet system. In: *CIA historical review program release as sanitized* [online]. 1999 Available at: https://www.cia.gov/library/readingroom/docs/DOC_0000497165.pdf [Accessed 20th January 2017].

[38] Weiss, G. W. (1996). The Farewell Dossier. Dumping the Soviets. *Studies in Intelligence. Central Intelligency Agency.* [Online] 39, 5. Avaible at: https://www.cia.gov/library/centr for the stud

225

y of intelligence/csi publications/csi studies/studies/9
6unclass/farewell.htm [Accessed 20th January 2017].

39 Erger, A. (2005). *Yoda and the Jedis: The Revolution in Military Affairs and the Transformation of War. The OST's Publication on Science & Technology Policy.* [Online] 7. Available at: http://www.ostina.org/content/view/274/ [Accessed 20th January 2017].

40 Hogan, H. (1967). *Lenin's Theory of Reflection. Master's Thesis.* McMaster University.

41 Dailey, B. D. and Parker P. (1987). *Soviet Strategic Deception.* Stanford: Hoover Institution Press.

42 Kasapoglu, C. (2015). Russia´s Renewed Military Thinking: Non-Linear Warfare and Reflective Control. *Research Paper. Rome: Research Division – NATO Defence College.* [Online] 121. Available at: http://www.ndc.nato.int/news/news.php?icode=877 [Accessed: 28th October 2017].

43 Rothstein, H., and Whaley, B. (2013). *The Art and Science of Military Deception (Artech House Intelligence and Information Operations).* New York: Artech House.

44Taham, S. (2013). *U.S. Governmental Information Operations and Strategic Communications: A Discredited Tool Or User Failure? : Implications for Future Conflict.* Carlisle Barracks, PA: Strategic Studies Institute, U.S. Army War College.

45 Either Dailey, B. D. and Parker P. (1987). *Soviet Strategic Deception.* Stanford: Hoover Institution Press. or Thomas, L. T. (2004). Russia's reflexive control theory and the military. *Journal of Slavic Military Studies*, 17.2, 237-256.

46 Dailey, B. D. and Parker P. (1987). *Soviet Strategic Deception.* Stanford: Hoover Institution Press.

47 Dailey, B. D. and Parker P. (1987). *Soviet Strategic Deception.* Stanford: Hoover Institution Press.
48 Thomas, L. T. (2004). Russia's reflexive control theory and the military. *Journal of Slavic Military Studies,* 17.2, pp. 237-256.
49 Operations Mincemeat, Fortitude as part of Bodyguard strategy
50 Reid, C. (1987) "Reflexive Control in Soviet Military Planning,". In Dailey, B. D. and Parker P. *Soviet Strategic Deception.* Stanford: Hoover Institution Press.
51 Reid, C. (1987) "Reflexive Control in Soviet Military Planning,". In Dailey, B. D. and Parker P. *Soviet Strategic Deception.* Stanford: Hoover Institution Press.
52 The origin of the prominence in the Soviet thinking is also based on the booklet of Lenin's *"What is to be done?"* where he laid out the importance of propaganda, agitation and political deception as integral elements of Communist party strategy. In O'Brien, T. N. (1989). *Russian Roulette: Disinformation in the U.S. Government and News Media. Master's Thesis.* South Carolina University Columbia.
53 Kasapoglu, C. (2015). Russia´s Renewed Military Thinking: Non-Linear Warfare and Reflective Control. *Research Paper. Rome: Research Division – NATO Defence College.* [Online] 121. Available at: http://www.ndc.nato.int/news/news.php?icode=877# [Accessed: 28th October 2017].
54 Kasapoglu, C. (2015). Russia´s Renewed Military Thinking: Non-Linear Warfare and Reflective Control. *Research Paper. Rome: Research Division – NATO Defence College.* [Online] 121. Available at:

http://www.ndc.nato.int/news/news.php?icode=877# [Accessed: 28th October 2017].

55 Kasapoglu, C. (2015). *Russia´s Renewed Military Thinking: Non-Linear Warfare and Reflective Control. Research Paper. Rome: Research Division – NATO Defence College.* [Online] 121. Available at: http://www.ndc.nato.int/ne ws/news.php?icode=877#[Accessed: 28th October 2017].

56 Hostile code and hostile content

57 Hostile code

58 Hostile code

59 Hostile content

60 Thomas, L. T. (2004). Russia's reflexive control theory and the military. *Journal of Slavic Military Studies*, 17.2, pp. 237-256.

61 Rothstein, H. and Whaley, B. (2013). *The Art and Science of Military Deception (Artech House Intelligence and Information Operations).* New York: Artech House.

62 Corman, S. R., and Dooley, K. J. (2009). *Strategic Communication on a Rugged Landscape. Principles for Finding the Right Message.* Consortium for Strategic Communication, Arizona State University. Tucson.

63 Thomas, L. T. (1996). Russian Views on Information-Based Warfare. *Airpower Journal*, Special. Edition, pp. 25-35.

64 Thomas, L. T. (2004). Russia's reflexive control theory and the military. *Journal of Slavic Military Studies*, 17.2, pp. 237-256.

65 Thomas, L. T. (2004). Russia's reflexive control theory and the military. *Journal of Slavic Military Studies*, 17.2, pp. 237-256.

[66] Kramer, D. F., Starr, S. and Wentz, H. L. (2009). *Cyberpower and National Security*. Dulles: Potomac Books, Inc.

[67] Gilbert, D. T. (1991). *How Mental Systems Believe*. American Psychologist. [Online] 46. Available at: http://www.danielgilbert.com/Gillber t%20(How%20Mental%20Systems%20Believe).PDF [Accessed: 28th January 2018].

[68] Panagiotis, O. (2016). *Strategic Military Deception. Prerequisites of Success in Technological Environment*. Available at: [Accessed: 28th January 2018].

[69] Panagiotis, O. (2016). *Strategic Military Deception. Prerequisites of Success in Technological Environment*. Available at: [Accessed: 28th January 2018].

[70] Panagiotis, O. (2016). *Strategic Military Deception. Prerequisites of Success in Technological Environment*. Available at: [Accessed: 28th January 2018].

[71] Korotchenko, Y. G. (1996). Information-Psychological Warfare in Modern Conditions. Military Thought. English edition. pp. 22-27.

[72] Komov, S. A. (1997). About Methods and Forms of Conducting Information Warfare. Military Thought. English edition. pp. 18-22.

[73] Ionov, M.D. (1995). On Reflective Enemy Control in a Military Conflict. Military Thought. English edition. pp. 45-50.

[74] Dailey, B. D. and Parker P. (1987). *Soviet Strategic Deception*. Stanford: Hoover Institution Press.

[75] Information security in the Russian perception represented by three layers in the previous chapter of this study.

[76] Thomas, Timothy. "Russia's reflexive control theory and the military." *Journal of Slavic Military Studies* 17.2 (2004): 237-256.

[77] Thomas, Timothy. "Russia's reflexive control theory and the military." *Journal of Slavic Military Studies* 17.2 (2004): 237-256.

[78] Fitzgerald, M. C. (1999). *Russian Views on IW, EW, and Command and Control: Implications for the 21st Century.* [Online]. Available at: [Accessed: 28th January 2018].

[79] Ionov, M.D. (1995). On Reflective Enemy Control in a Military Conflict. Military Thought. English edition. p. 45-50.

[80] Komov, S. A. (1997). About Methods and Forms of Conducting Information Warfare. Military Thought. English edition. pp. 18-22.

[81] General Valery Gerasimov, the Chief of General Staff emerged in *Voyenno- Promyshlennyy Kuryer* In McDermott, R. (2014). *Gerasimov Unveils Russia's 'Reformed' General Staff. Eurasia.* Daily Monitor Volume: 11 Issue: 27. [Online]. Available at: https://jamestown.org/program/gerasimov-unveils-russias-reformed-general-staff/ [Accessed: 28th January 2018].

[82] Russian Federation. President of the Russian Federation.(2000). *Information Security Doctrine of the Russian Federation.* [Online]. Available at: https://info.publicintelligence.net/RU-InformationSecurity-2000.pdf [Accessed 20th January 2017].

[83] Russian Federation. President of the Russian Federation (2015). *Doctrine of information Security of the Russian Federation.* [Online]. Available at: http://www.mid.ru/en/foreign_policy/official_documents/-

/asset_publisher/CptICkB6BZ29/content/id/2563163 [Accessed 20th January 2017].

84Russian Federation. President of the Russian Federation 2015). *Russian Federation National Security Strategy.* [Online]. Available atfrom: http://www.ieee.es/Galerias/fichero/OtrasPublicaciones/Internacional/2016/Russian-National-Security-Strategy-31Dec2015.pdf [Accessed 20th January 2017].

85 Russian Federation. President of the Russian Federation (2010). *Military doctrine of the Russian Federation.* [Online]. Available at:http://carnegieendowment.org/files/2010russia_military_doctrine.pdf [Accessed 20th January 2017].

86 Russian Federation. President of the Russian Federation (2010). *Russian Federation Armed Forces' Information Space Activities Concept.* [Online]. Available at: http://eng.mil.ru/en/science/publications/more.htm?id=10845074@cmsArticle [Accessed 20th January 2017].

87 Military Review, (2016). *The Value of Science Is in the Foresight New Challenges Demand Rethinking the Forms and Methods of Carrying out Combat Operations.* [Online]. Available at: https://usacac.army.mil/CAC2/MilitaryReview/Archives/English/MilitaryReview_20160228_art008.pdf [Accessed 18th March 2018].

88 Russian Federation. President of the Russian Federation (2014). *The Military Doctrine of the Russian Federation.* [Online]. Available at: https://rusemb.org.uk/press/2029 [Accessed 20th January 2017].

89 Russian Federation. President of the Russian Federation (2016). *Doctrine of Information Security of the Russian Federation* [Online]. Available at:

http://www.mid.ru/en/foreign_policy/official_docum
ents/ /asset_publisher/CptICkB6BZ29/content/id/25
63163 [Accessed 20th January 2017].

90 The New York Times, (2015). *The Agency.*
[Online]. Available at:
https://www.nytimes.com/2015/06/07/magazine/the-
agency.html [Accessed 18th March 2018].

91 President Trump signs into law U.S. government
ban on Kaspersky Lab software. *Reuters* [online].
Available at: https://www.reuters.com/article/us-usa-
cyber-kaspersky/trump-signs-into-law-u-s-
government-ban-on-kaspersky-lab-software-
idUSKBN1E62V4 [Accessed 18th March 2018].
https://www.reuters.com/article/us-usa-cyber-
kaspersky/trump-signs-into-law-u-s-government-ban-
on-kaspersky-lab-software-idUSKBN1E62V4

92 US slaps China's ZTE with 7-year components
ban for breaching terms of sanctions settlement. *South
China morning post* [online]. Available at:
httpshttp://www.scmp.com/business/companies/artic
le/2142002/us-slaps-zte-seven-year-components-ban-
breaching-terms-sanctions[Accessed 18th March
2018].

93 Washington Post, (2014). Putin says Russia will
protect the rights of Russian
abroad. [Online]. Available at: https://www.washingto
npost.com/world/transcript-putin-says-russia-will-
protect-the-rights-of-russians-
abroad/2014/03/18/432a1e60-ae99-11e3-a49e-
76adc9210f19_story.html?utm_term=.6d532d8be341
[Accessed 18th March 2018]

94 NATO Cooperative Cyber Defence Centre of
Excellence, (2015). *An Updated Draft of the Code of
Conduct Distributed in the United Nations – What's
New?* [Online]. Available at:

https://ccdcoe.org/updated-draft-code-conduct-distributed-united-nations-whats-new.html#footnote3_7g6pxbk [Accessed 18th March 2018].

95 See, e.g.: Carr, J. and Dao, ´D. (2011). *"4 Problems with China and Russia´s International Code of Conduct for Information Security"*, and Dao,´ D. and Giles, K. 2011). *"Russia´s Public Stance on Cyberspace Issues"*, in Czosseck, Ch. Ottis, F. And Ziolkowski, K. (eds.) (2012). 4th International Conference on Cyber Conflict. Tallin: NATO CCD COE Publications.

96 Such as the the news about NATO CCD CoE involvement with the website Mirotvorce, to be described later in the book.

97 See chapter "Cyber power in Russia"

98 Weiss, G. W. (1996). The Farewell Dossier. Dumping the Soviets. *Studies in Intelligence. Central Intelligency Agency.* [Online] 39, 5. Avaiilible at: https://www.cia.gov/library/centr for the stud y of intelligence/csi publications/csi studies/studies/96unclass/farewell.htm [Accessed 20th January 2017].

99 EFJ, (2017). *Russian 'foreign agents' media law threatens media freedom.* [Online]. Available at: https://europeanjournalists.org/blog/2017/11/28/russian-foreign-agents-media-law-threatens-media-freedom/ [Accessed 18th March 2018].

100 VOA News, (2018). *Russia's Foreign Agent Law Has Chilling Effect On Civil Society Groups, NGOs.* [Online]. Available at: https://www.voanews.com/a/russia-labels-media-outlets-as-foreign-agents/4221609.html [Accessed 18th March 2018].

101 Interestingly, Russian agents of influence circumvent the U.S. FARA legislation by registering themselves as NGOs. In: MCFaul, M. (2017) 6th

January. Available at https://twitter.com/McFaul [(Accessed: 18th March 2017].

102 The Moscow Times, (2014). *Vkontakte Founder Says Sold Shares Due to FSB Pressure.* [Online]. Available at: https://themoscowtimes.com/news/vkontakte-founder-says-sold-shares-due-to-fsb-pressure-34132 [Accessed 18th March 2018]. And Business Insider, (2014). and The Moscow Times, (2014). *Putin Has Taken Control Of Russian Facebook.* [Online]. Available at: http://www.businessinsider.com/putin-has-taken-control-of-russian-facebook-2014-4 [Accessed 18th March 2018].,

103 SORM or Systema Operativno -Razisknikh Meropriatiy – the System of Operative -Search Measures on Communications (comms interception) In: Soldatov, A. (2014). Russia's communications interception practices (SORM) [presentation] *Ag entu-ra.Ru.* [Online]. Available at: http://www.europarl.eur opa.eu/meetdocs/2009_2014/documents/libe/dv/sol datov_presentation_/soldatov_presentation_en.pdf [Accessed: 7th June 2017].

104Created by a presidential decree In: [Online]. Ava ilible at: http://static.kremlin.ru/media/events/files/4 1d4a95e0e2d01da1117.pdf [Accessed 18th March 2018].

105 RT being an organization of strategic importance to the Russian Federation according to official list of core organizations. In: *The list of backbone organizations, approved by the Government Commission to improve the sustainability of the development of the Russian econo-my* [Online]. Available at: https://web.archive.org/we b/20081227071316/http:/www.government.ru/conten

t/governmentactivity/mainnews/33281de212bf49fdbf3
9d611cadbae95.doc [Accessed 18th March 2018].

[106]In case you weren't clear on Russia Today's relationship to Moscow, Putin clears it up. *The Washington Post* [online]. Available at: https://www.washingtonpost.com/news/world views/wp/2013/06/13/in-case-you-werent-clear-on-russia-todays-relationship-to-moscow-putin-clears-it-up/?noredirect=on&utm_term=.9be5f64c0a34 [Accessed 18th March 2018].ab3a708f83d4

[107] Radio Free Europe Radio Liberty, (2015). *Russian TV Deserters Divulge Details On Kremlin's Ukraine 'Propaganda'* [Online]. Available at: https://www.rferl.org/a/russian-television-whistleblowers-kremlin-propaganda/27178109.html [Accessed 18th March 2018].

[108] Human Rights First, (2017). *Russian Influence in Europe. [Online]*. Available at: https://www.humanrightsfirst.org/resource/russian-influence-europe [Accessed 18th March 2018].

[109] NATO Cooperative Cyber Defence Centre of Excellence, (2015). *An Updated Draft of the Code of Conduct Distributed in the United Nations – What's New?* [Online]. Available at: https://ccdcoe.org/updated-draft-code-conduct-distributed-united-nations-whats-new.html#footnote3_7g6pxbk [Accessed 18th March 2018].

[110] DW News, (2018). *Russia moves toward creation of an independent internet.* [Online]. Available at: http://www.dw.com/en/russia-moves-toward-creation-of-an independent-internet/a-42172902 [Accessed 18th March 2018].

[111] Medvedev, S. A. (2015). *Offense-defense theory analysis of Russian cyber capability.* PhD Thesis. Naval Postgraduate School.

[112] Reuters, (2017). Russia's RT America registers as 'foreign agent' in U.S. [Online]. Available at: https://www.reuters.com/article/us-russia-usa-media-restrictions-rt/russias-rt-america-registers-as-foreign-agent-in-u-s-idUSKBN1DD25B [Accessed 18th March 2018].

[113] For example the cases of Alexandr Kuranov and Vladimir Snergirev.

[114] Thomas, L. T. (1996). Russian Views on Information-Based Warfare. *Airpower Journal,* Special. Edition, pp. 25-35.

[115] Fink, A. L. (2017). *The Evolving Russian Concept of Strategic Deterrence: Risks and Responses.* Arms Control Association. [Online]. Availa ble at: https://www.armscontrol.org/act/2017-07/features/evolving-russian-concept-strategic-deterrence-risks-responses [Accessed: 28th January 2018]

[116] Bruusgaard, Ven K. (2016). *Global Politics and Strategy August–September 2016.* IISS. Survival. Global Politics and Strategy, [Online]. Availab le at: https://www.iiss.org/en/publications/survival/se ctions/2016-5e13/survival--global-politics-and-strategy-august-september-2016-2d3c/58-4-02-ven-bruusgaard-45ec [Accessed: 7th June 2016].

[117] DW: Russia moves toward creation of an independent internet [online]. Available at: https://www.dw.com/en/russia-moves-toward-creation-of-an-independent-internet/a-42172902 [Accessed 20th January 2017

[118] [online]. Available at: http://www.ieee.es/Galerias/fichero/OtrasPublicacion

es/Internacional/2016/Russian-National-Security-Strategy-31Dec2015.pdf [Accessed 20th January 2017]

[119] Russian Federation. President of the Russian Federation 2015). *Russian Federation National Security Strategy.* [Online]. Available at: http://ww w.ieee.es/Galerias/fichero/OtrasPublicaciones/Interna cional/2016/Russian-National-Security-Strategy-31Dec2015.pdf [Accessed 20th January 2017].

[120] Russian Federation. President of the Russian Federation (2010). Military doctrine of the Russian Federation. [Online]. Available at:http://carnegieendowment.org/files/2010russia_mi litary_doctrine.pdf [Accessed 20th January 2017].pdf

[121] *"The Military Doctrine reflects the Russian Federation's adherence to the utilization of political, diplomatic, legal, economic, environmental, **informational**, military, and other instruments for the protection of the national interests of the Russian Federation and the interests of its allies."*

[122] Russian Federation. President of the Russian Federation (2010). *Military doctrine of the Russian Federation.* [Online]. Available at:http://c arnegieendow-ment.org/files/2010russia_military_doctrine.pdf [Accessed 20th January 2017].

[123] Russian Federation. President of the Russian Federation (2010). *Military doctrine of the Russian Federation.* [Online]. Available at:http://c arnegieendow-ment.org/files/2010russia_military_doctrine.pdf [Accessed 20th January 2017].

[124] Russian Federation. President of the Russian Federation (2010). *Military doctrine of the Russian Federation.* [Online]. Available at:

http://carnegieendowment.org/files/2010russia_milit
ary_doctrine.pdf [Accessed 20th January 2017].

125 Russian Federation. President of the Russian
Federation (2010). *Military doctrine of
the Russian Federation.* [Online]. Available at:
http://carnegieendowment.org/files/2010russia_milit
ary_doctrine.pdf [Accessed 20th January 2017].

126 Medvedev, S. A. (2015). *Offense-defense theory
analysis of Russian cyber capability.* PhD Thesis.
Naval Postgraduate School.

127 [online]. Available at: https://icds.ee/natos-
cyber-defence-after-warsaw/Vyhlaseni [Accessed
20th January 2017].

128 Ministry of Defence of the Russian Federation
(2000). *Russian Federation Armed Forces' Information Space Activities Concept.* [Online]. Available at:
http://eng.mil.ru/en/science/publications/more.htm?i
d=10845074@cmsArticle [Accessed 20th January
2017].

129 Ministry of Defence of the Russian Federation
(2000). *Russian Federation Armed Forces' Information Space Activities Concept.* [Online]. Available at:
http://eng.mil.ru/en/science/publications/more.htm?i
d=10845074@cmsArticle [Accessed 20th January
2017].

130 Ministry of Defence of the Russian Federation
(2000). *Russian Federation Armed Forces' Information Space Activities Concept.* [Online]. Available at:
http://eng.mil.ru/en/science/publications/more.htm?i
d=10845074@cmsArticle [Accessed 20th January
2017].

131 Ministry of Defence of the Russian Federation (2000). *Russian Federation Armed Forces' Information Space Activities Concept*. [Online]. Available at: http://eng.mil.ru/en/science/publications/more.htm?id=10845074@cmsArticle [Accessed 20th January 2017].

132 Ministry of Defence of the Russian Federation (2000). *Russian Federation Armed Forces' Information Space Activities Concept*. [Online]. Available at: http://eng.mil.ru/en/science/publications/more.htm?id=10845074@cmsArticle [Accessed 20th January 2017].

133 Ministry of Defence of the Russian Federation (2000). *Russian Federation Armed Forces' Information Space Activities Concept*. [Online]. Available at: http://eng.mil.ru/en/science/publications/more.htm?id=10845074@cmsArticle [Accessed 20th January 2017].

134 General Valery Gerasimov, the Chief of General Staff emerged in *Voyenno- Promyshlennyy Kuryer* In McDermott, R. (2014). *Gerasimov Unveils Russia's 'Reformed' General Staff. Eurasia.* Daily Monitor Volume: 11 Issue: 27. [Online].

135 General Valery Gerasimov, the Chief of General Staff emerged in *Voyenno- Promyshlennyy Kuryer* In McDermott, R. (2014). *Gerasimov Unveils Russia's 'Reformed' General Staff. Eurasia.* Daily Monitor Volume: 11 Issue: 27. [Online].

136 General Valery Gerasimov, the Chief of General Staff emerged in *Voyenno- Promyshlennyy Kuryer* In McDermott, R. (2014). *Gerasimov Unveils Russia's*

'Reformed' General Staff. *Eurasia*. Daily Monitor Volume: 11 Issue: 27. [Online].

[138] Russian Federation. President of the Russian Federation (2014). *The Military Doctrine of the Russian Federation*. [Online]. Availabl e at: https://rusemb.org.uk/press/2029 [Accessed 20th January 2017].

[139] Worldview Stratfor: The Future of Russia's Military: Part 1 [online]. Available at: https://worldview.stratfor.com/article/future-russias-military-part-1 [Accessed: 28th January 2018]

[140]Tatham, S. (2013). *U.S. Governmental Information Opera tions and Strategic Communications: A Discredited Tool Or User Failure? : Implications for Future Conflict*. Carlisle Barracks, PA: Strategic Studies Institute, U.S. Army War College.

[141] Crutcher, M. H. (2000). *The Russian armed forces at the dawn of the millennium*. Carlisle: U.S. Army War College Carlisle and Thomas, Timothy L. (2000). *The Russian View Of Information War*. In The Russian Armed Forces at the Dawn of the Millenium. [Online]. Available at: *www.dtic.mil/dtic/tr/fulltext/u2/a423593.pdf* [(Accessed: 28th January 2018].

[142]Tatham, S. (2013). *U.S. Governmental Information Opera tions and Strategic Communications: A Discredited Tool Or User Failure? : Implications for Future Conflict*. Carlisle Barracks, PA: Strategic Studies Institute, U.S. Army War College.

[143] Crutcher, M. H. (2000). *The Russian armed forces at the dawn of the millennium*. Carlisle: U.S. Army War College Carlisle and Thomas, Timothy L. (2000). *The Russian View Of Information War*. In The

Russian Armed Forces at the Dawn of the Millenium. [Online]. Available at: *www.dtic.mil/dtic/tr/fulltext/u2/a423593.pdf* [Accessed: 28th January 2018].

144 Thomas, L. T. (2004). Russia's reflexive control theory and the military. *Journal of Slavic Military Studies*, 17.2, pp. 237-256.

145Komov, S. A. (1997). About Methods and Forms of Conducting Information Warfare. *Military Thought*, 4, pp. 18-22.

146 Komov, S. A. (1997). About Methods and Forms of Conducting Information Warfare. *Military Thought*, 4, pp. 18-22.

147 Komov, S. A. (1997). About Methods and Forms of Conducting Information Warfare. *Military Thought*, 4, pp. 18-22.

148 FitzGerald, M. C. (1999). *Russian Views on IW, EW, and Command and Control: Implications for the 21st Century.* [Online]. Available at: http://www.dodccrp.org/events/1999_CCRTS/pdf_fil es/track_5/089fitzg.pdf [Accessed: 28th January 2018].

149Thomas, L. T. (2004). Russia's reflexive control theory and the military. *Journal of Slavic Military Studies*, 17.2, pp. 237-256.

150 L´Express, (2015). *Piratage de TV5 Monde: l'enquête s'oriente vers la piste russe.* [Online]. Available at: https://www.lexpress.fr/a ctualite/medias/piratage-de-tv5-monde-la-piste-russe_1687673.html [Accessed 18 March 2018].

151 Trend Micro, (2016). *Operation Pawn Storm: Fast Facts and the Latest Developments.* [Online]. Available at: https://www.tre ndmicro.com/vinfo/us/security/news/cyber-

attacks/operation-pawn-storm-fast-facts [Accessed 18 March 2018].

152 Denial of Service by overload of requests.

153 Distributed denial of service using a network of devices with computational power to overload the capacity of the information sources, this case web servers hosting government information portals.

154 Komov, S. A. (1997). About Methods and Forms of Conducting Information Warfare. *Military Thought*, 4, pp. 18-22.

155 This information was obtained during a closed door interview with a high ranking law enforcement official from Ukraine with direct knowledge and understanding of ongoing cases and cyber security threats towards Ukraine posed by Russia.

156 RT, (2015). *NATO trace 'found' behind witch hunt website in Ukraine*. [Online]. Available at: https://www.rt.com/news/253117-nato-ukraine-terrorsite/ [Accessed 15th February 2018].

157 NATO Cooperative Cyber Defence Centre of Excellence. [Online]. Available at: https://ccdcoe.org/index. html [Accessed 18th March 2018].

160 VOA NEWS, (2017). *Sinister Text Messages Reveal High tech Front in Ukraine War*. [Online]. Available at: https://www.voanews.com/a/sinister-text-messages-high tech-frony-ukraine-war/3848034.html [Accessed 18th March 2018].

161 Russian Military Review, (2015). *День инноваций ЮВО: комплекс РЭБ РБ-341В «Леер-3»*. [Online]. Available at: https://archive.is/MDecB [Accessed 18th March 2018].

162 Inform Napalm, (2016). *Russian Leer-3 EW system revealed in Donbas*. [Online]. Available at: https://informnapalm.org/en/russian-leer-3wf-donbas/ [Accessed 18th March 2018].

[163] Based on a private conversation with a high ranking military officer, a general officer, close to the chairman of the NATO country's General Staff, fell for the story.

[164] Digital Forensic Research Lab, (2017). *Russia's Fake "Electronic Bomb". How a fake based on a parody spread to the Western mainstream.* [Online]. Available at: https://medium.com/dfrlab/russias-fake-electronic-bomb-4ce9dbbc57f8 [Accessed 18th March 2018].

[165] U. S. Department of Defense, (2014). *Russian Aircraft Flies Near U.S. Navy Ship in Black Sea.* [Online]. Available at: http://archive.defens e.gov/news/newsarticle.aspx?id=122052 [Accessed 18th March 2018].

[166] Digital Forensic Research Lab, (2017). *Russia's Fake "Electronic Bomb". How a fake based on a parody spread to the Western mainstream.* [Online]. Available at: https://medium.com/dfrlab/ru ssias-fake-electronic-bomb-4ce9dbbc57f8 [Accessed 18th March 2018].

[167] Sputnik News (2014). *Russische SU-24 legt amerikanischen Zerstörer lahm.* [Online]. Available at: https://de.sputniknews.com/meinungen/2014042126 8324381-Russische-SU-24-legt-amerikanischen-Zerstrer-lahm/ [Accessed 18th March 2018].

[168] RG RU, (2016). *Что напугало американский эсминец.* [Online]. Available at: https://rg.ru/2014/04/30/reb-site.html [Accessed 18th March 2018].

[169] Voltaire Network, (2016). *About Voltaire Network.* [Online]. Available at: http://www.voltairenet.org/article150341.html [Accessed 18th March 2018].

[174] Info Wars. [Online]. Available at: https://www.infowars.com/ [Accessed 18th March 2018].

[175] Info Wars, (2014). *Russians Disable U.S. Guided Missile Destroyer.* [Online]. Available at: https://www.infowars.com/russians-disable-u-s-guided-missile-destroyer/ [Accessed18th March 2018].

[178] Concern Radio Electronic Technologies. [Online]. Available at: http://rostec.ru/en/about/companies/346/ [Accessed 18th March 2018].

[179] NDE7, (2015). Russian EW-technologies are among the most advanced in the Word. [Online]. Available at: http://archive.is/NDE7r#selection 837.0 839.1 [Accessed 18th March 2018].

[180] Digital Forensic Research Lab, (2017). *Russia's Fake "Electronic Bomb". How a fake based on a parody spread to the Western mainstream.* [Online]. Available at: https://medium.com/dfrlab/russias-fake-electronic-bomb-4ce9dbbc57f8 [Accessed 18th March 2018].

[181] Fabricated comments by leading personalities in the field have the aim of selling the story to people and veil their suspicion.

[183] Vesti News, (2017). *Electronic Warfare: How to Neutralize the Enemy Without a SingleShot.* [Online] Available at: https://www.youtube.com/watch?v=vI4uS307ydk&feature=youtu.be [Accessed 18 March 2018].

[184] Digital Forensic Research Lab, (2017). *Russia's Fake "Electronic Bomb". How a fake based on a parody spread to the Western mainstream.* [Online]. Available at: https://medium.com/dfrlab/russias-fake-electronic-bomb-4ce9dbbc57f8 [Accessed 18th March 2018].

[185]Washington Post, (2017). *Fingered for Trading in Russian Fake News.* [Online]. Available at: https://www.washingtonpost.com/blogs/erik wemple/wp/2 017/06/07/foxnews-com-fingered-for-trading-in-russian-fake news/?utm_term=.647f82ed21dc [Accessed 18th March 2018].

[187] United States. Dept. of State (1989). *Soviet influence activities: a report on active measures and propaganda, 1987-1988.* New York: U.S. Dept. of State.

[188] Kragh, M. and Åsberg, S. (2017). Russia's strategy for influence through public diplomacy and active measures: the Swedish case. *Journal of Strategic Studies,* 40.6, pp. 773-816.

[189] CyberBerkut (2014). *CyberBerkut hacked Kiev digital billboards.* [Online]. Available at: https://cyber-berkut.org/en/olden/index2.php [Accessed 20th January 2017].

[190] Украина Сегодня, (2015). *'Cyberberkut' hacked Kyiv billboards* . [Online]. Available at: https://www.youtube.com/watch?v=E8A2MIkiavE [Accessed 18th March 2018].

[192] DEF CON (also written as DEFCON, Defcon, or DC), one of the world's largest hacker conventions. [Online]. Available at: https://www.defcon.org/ [Accessed 20th January 2017].

[193] Tottenkoph, R. (2016). Hijacking the Outdoor Digital Billboard [PowerPoint presentation]. *DEF CON Hacking Confe rence.* [Online]. Available at: https://www.defcon.org/images/defcon-16/dc16-presentations/defcon-16-tottenkoph-rev-philosopher.pdf [Accessed: 7th June 2016].

[194] CyberBerkut (2014). *E-mail of the Ukrainian Ministry of Defense colonel has been hacked* [Online]. Available atfrom: https://cyber-berkut.org/en/olden/index2.php [Accessed 20th January 2017].

[195] CyberBerkut (2014). *New punishers' losses data in the South-East.* [Online]. Available at: https://cyber-berkut.org/en/olden/index2.php [Accessed 20th January 2017].

[196] Украина Сегодня, (2015). *'Cyberberkut' hacked Kyiv billboards.* [Online]. Available at: https://www.youtube.com/watch?v=E8A2MIkiavE [Accessed 18th March 2018].

[197] Shoulder-mounted missile launcher

[198] Logvinov, A. (2015). *Ukrainian rebels make fake video using weapons from the game 'Battlefield 3'.* Meduza. [Online]. Available at: https://meduza.io/en/lion/2015/07/23/ukrainian-rebels-make-fake-video-using-weapons-from-the-game-battlefield-3 [Accessed: 28th January 2018].

[199] Thomas, L. T. (2004). Russia's reflexive control theory and the military. *Journal of Slavic Military Studies*, 17.2, pp. 237-256.

[200] CyberBerkut (2014). *CyberBerkut suspended the operation of the Ukrainian Central Election Commission* [Online]. Available at: https://cyber-berkut.org/en/olden/index2.php [Accessed 20th January 2017].

[201] Fitzgerald, M. C. (1999). *Russian Views on IW, EW, and Command and Control: Implications for the 21st Century.* [Online]. Available at: http://www.dodccrp.org/events/1999_CCRTS/pdf_files/track_5/089fitzg.pdf [Accessed: 28th January 2018].

[202] Fitzgerald, M. C. (1999). *Russian Views on IW, EW, and Command and Control: Implications for the 21st Century.* [Online]. Available at: http://www.dodccrp.org/events/1999_CCRTS/pdf_fil es/track_5/089fitzg.pdf [Accessed: 28th January 2018].

[203] Fitzgerald, M. C. (1999). *Russian Views on IW, EW, and Command and Control: Implications for the 21st Century.* [Online]. Available at: http://www.dodccrp.org/events/1999_CCRTS/pdf_fil es/track_5/089fitzg.pdf [Accessed: 28th January 2018].

[204] E.g. technological, satellites, communication lines, or in case of psychological targeting the human element of decision-making cycle.

[205] Fitzgerald, M. C. (1999). *Russian Views on IW, EW, and Command and Control: Implications for the 21st Century.* [Online]. Available at: http://www.dodccrp.org/events/1999_CCRTS/pdf_fil es/track_5/089fitzg.pdf [Accessed: 28th January 2018].

[206] Fitzgerald, M. C. (1999). *Russian Views on IW, EW, and Command and Control: Implications for the 21st Century.* [Online]. Available at: http://www.dodccrp.org/events/1999_CCRTS/pdf_fil es/track_5/089fitzg.pdf [Accessed: 28th January 2018].

[207] Fitzgerald, M. C. (1999). *Russian Views on IW, EW, and Command and Control: Implications for the 21st Century.* [Online]. Available at: http://www.dodccrp.org/events/1999_CCRTS/pdf_fil es/track_5/089fitzg.pdf [Accessed: 28th January 2018].

[208] Key takeaways from an event "Understanding Russian Strategic Behavior Project", Research Seminar, George C. Marshall Center.

[209] According to Chatham House rule, the author can not be disclosed or the content attributed.

[210] Calhoun, L. (2018). Elon Musk Just Sent More Stuff Into Space--This Time, It's Even Better Than the Roadster. *Inc.* [Online]. Available at: [Accessed: 28th January 2018].

[211] Fitzgerald, M. C. (1999). *Russian Views on IW, EW, and Command and Control: Implications for the 21st Century.* [Online]. Available at: http://www.dodccrp.org/events/1999_CCRTS/pdf_fil es/track_5/089fitzg.pdf [Accessed: 28th January 2018].

[212] Godson, R. (2017). *Written Testimony of ROY GODSON to the Senate Select Committee on Intelligence, Open Hearing, March 30, 2017 "Disinformation: A Primer in Russian Active Measures and Influence Campaigns."* [Online]. Available at: https://www.intelligence.senate.gov/sites/default/files /documents/os-rgodson-033017.pdf [Accessed: 28th January 2018].

[213] Godson, R. (2017). *Written Testimony of ROY GODSON to the Senate Select Committee on Intelligence, Open Hearing, March 30, 2017 "Disinformation: A Primer in Russian Active Measures and Influence Campaigns."* [Online]. Available at: https://www.intelligence.senate.gov/sites/default/files /documents/os-rgodson-033017.pdf [Accessed: 28th January 2018].

[214] Political joke on Lenin.

[215] Lenin, V. I. (1902). *What Is To Be Done?*. Marxists Internet Archive. [Online]. Available at https://www.marxists.org/archive/lenin/works/download/what-itd.pdf [Accessed: 28th January 2018].

[216] Schoen, F. and Lamb, C. (2012). *Deception, disinformation, and strategic communications*. Washington D.C.: National Defense University Press.

[217] Beaumont, R. (1982). *Maskirovka: Soviet Camouflage, Concealment and Deception*. College Station, Texas: Center for Strategic Technology, A & M University System.

[218] Rothstein, H., and Whaley, B. (2013). *The Art and Science of Military Deception (Artech House Intelligence and Information Operations)*. New York: Artech House.

[219] Russia's renewed military thinking: Non-linear warfare and reflexive control. National Defense University [online]. Available at: http://cco.ndu.edu/Portals/96/Documents/Articles/russia%27s%20renewed%20Military%20Thinking.pdf [Accessed 20th January 2017].

[220] Russia's renewed military thinking: Non-linear warfare and reflexive control. National Defense University [online]. Available at: http://cco.ndu.edu/Portals/96/Documents/Articles/russia%27s%20renewed%20Military%20Thinking.pdf [Accessed 20th January 2017].

[221] Russia's renewed military thinking: Non-linear warfare and reflexive control. National Defense University [online]. Available at: http://cco.ndu.edu/Portals/96/Documents/Articles/russia%27s%20renewed%20Military%20Thinking.pdf [Accessed 20th January 2017].

[222] Russia's renewed military thinking: Non-linear warfare and reflexive control. National Defense University [online]. Available at: http://cco.ndu.edu/Portals/96/Documents/Articles/russia%27s%20renewed%20Military%20Thinking.pdf [Accessed 20th January 2017].

[223] Fitzgerald, M. C. (1999). *Russian Views on IW, EW, and Command and Control: Implications for the 21st Century.* [Online]. Available at: http://www.dodccrp.org/events/1999_CCRTS/pdf_files/track_5/089fitzg.pdf [Accessed: 28th January 2018].

[224] Fitzgerald, M. C. (1999). *Russian Views on IW, EW, and Command and Control: Implications for the 21st Century.* [Online]. Available at: http://www.dodccrp.org/events/1999_CCRTS/pdf_files/track_5/089fitzg.pdf [Accessed: 28th January 2018].

Daniel P. Bagge is the Cyber Attaché of the Czech Republic to the United States and Canada. He works for the National Cyber and Information Security Agency (NCISA), and is based at the Embassy of the Czech Republic in Washington D.C. Previously he was the Director of Cyber Security Policies at NCISA. Being responsible for the implementation of the National Cyber Security Strategy he co-authored, he established the Strategic Information and Analysis, Education and Exercise, International Organizations and Law, National Strategies and Policies and the Critical Information Infrastructure Protection Units within the National Cyber Security Center. He provided the expertise of his department to ACT NATO, USCYBERCOM, USAFRICOM, the U.S. Congress, Military and National Security entities in Ukraine, the Balkans and other states. He holds an MA from the Bundeswehr University in Munich / George C. Marshall Center in Germany.

Made in the USA
San Bernardino, CA
09 August 2019